IMAGES
of America

PORTSMOUTH FIREFIGHTING

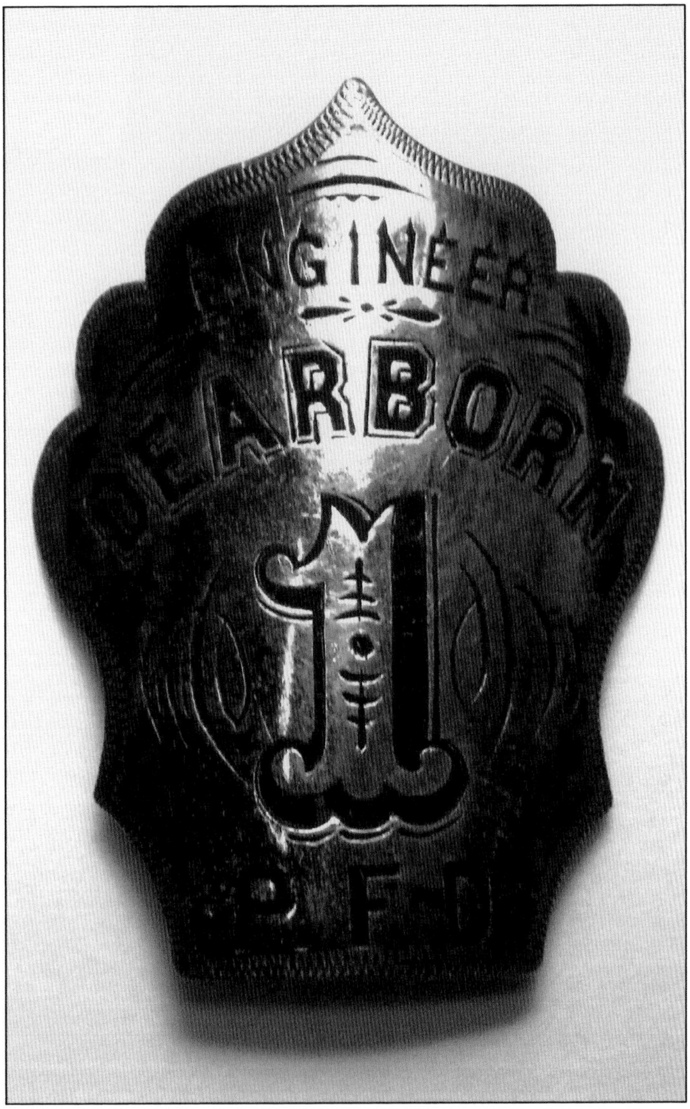

On June 30, 1865, the Portsmouth Fire Department received a steam fire engine built by Campbell, Whittier and Company of Roxbury, Massachusetts. Originally slated to be named the Portsmouth, the fire engine was named for city mayor and former fireman Jonathan Dearborn. The Dearborn Steam Fire Engine Company No. 1 consisted of 26 hook and ladder men that ran out of the Central engine house on Court Street from 1865 to 1876. It was then placed in reserve at the Hanover Street firehouse. This engineer's badge is one of the few remaining artifacts from Portsmouth's second steam fire engine company.

On the cover: Members of Portsmouth's Willard J. Sampson Hook and Ladder Company No. 1 make a fine appearance in this image from the late 1880s. This photograph was taken during one of the many Firemen's Day annual parades, when great throngs of citizens would line the streets of Portsmouth looking forward to the brilliant and inspiring spectacle of marching firemen and their beautifully decorated fire apparatus. It was reported after one such event that no city could show a finer-looking body of "fire laddies," nor one representing more glorious traditions. (Courtesy of Strawbery Banke.)

IMAGES
of America

PORTSMOUTH FIREFIGHTING

Steven E. Achilles

ARCADIA
PUBLISHING

ISBN 978-0-7385-6210-0

Published by Arcadia Publishing
Charleston, South Carolina

Printed in the United States of America

Library of Congress Catalog Card Number: 2008934666

For all general information contact Arcadia Publishing at:
Telephone 843-853-2070
Fax 843-853-0044
E-mail sales@arcadiapublishing.com
For customer service and orders:
Toll-Free 1-888-313-2665

Visit us on the Internet at www.arcadiapublishing.com

For the daughters and sons of firefighters

Contents

ACKNOWLEDGMENTS

I am very thankful for the contributions, time, and sincere interest of the following: the special collections staff of the Portsmouth Public Library; Tara Webber, collections manager, Strawbery Banke Museum; Robert Randall of Kittery, Maine; Judi (Connors) O'Brien and John Connors of Portsmouth; past members of the Kearsarge Steam Fire Engine Historical Association; retired deputy chief George Pierce and retired deputy chief Eugene Hersey of the Portsmouth Fire Department; Portsmouth deputy fire chief Steve Griswold; and Erin Rocha, my editor at Arcadia Publishing.

I am also grateful for the advice, support, and understanding given to me by my wife, Margaret, and daughters Emily and Katie as I researched and arranged the stories, photographs, and historical facts for this book.

Unless otherwise noted, all images appearing in this book are from the Portsmouth Fire Department photograph collection.

INTRODUCTION

Settled in 1623, Portsmouth is situated along the shores of the Piscataqua River between New Hampshire and Maine and lies just five miles upstream from the Atlantic Ocean. Initially known as Strawbery Banke, this waterfront community thrived as a center of Colonial maritime commerce. With no organized fire services in existence, "night watches" patrolled the streets with the mission of discovering fires. When a fire was discovered, the watchman in the area would cry out the location. Church bells would be rung to alert the community and to summon aid. All able-bodied citizens would then run to the scene with fire buckets.

At the March 26, 1744, town meeting, a committee was established to purchase a "New Ingine." Twelve years later, on September 13, 1756, a special town meeting was called to consider the best measures to extinguish and prevent the spread of fire and to regulate the labor and proceeding of people assembled on such occasions. In addition to agreeing to purchase fire tools and buckets, 12 "firewards" were elected to oversee fire protection in the town. In 1781, the first great fire in Portsmouth broke out in a barn standing on the site of the present Music Hall, consuming, among other buildings, the jail and Gov. John Langdon's house.

Throughout the latter part of the 18th century, the people of Portsmouth continued to improve their fire protection services through the purchase of fire engines and the building of engine houses. Providing support to the citizen volunteer firefighters and firewards were members of the many Portsmouth fire societies. The primary purpose of these fire societies was not to extinguish fires but to salvage and protect property recovered from a fire in a member's home or business. The tools carried by each member usually consisted of two leather buckets, a bed wrench, a large four-bushel cloth bag for carrying valuables, and a long-handled mop. Portsmouth's six fire societies were the United (1761), Friendly (1762), Federal (1789), Humane (1796), Mechanic (1811), and Alert (1813).

At the beginning of the 19th century, Portsmouth experienced three notable fires that would forever shape the city. The first was on December 25, 1802, in which 132 buildings were destroyed. On December 24, 1806, the Chapel Street fire broke out, leveling St. John's Church and 13 other buildings. The last and the most devastating happened on December 22, 1813, when over 270 buildings from Court Street to the river were destroyed. As a result, the citizens voted to prohibit the erection of wooden buildings over 12 feet high in the town. The act passed, and today a majority of the city's downtown is still masonry buildings with slate roofs. Portsmouth in 1839 was divided into five fire districts with engine companies assigned to each district. These companies were under the direction of the firewards, who annually chose two directors from their ranks to command each engine company. Fire company members were required to be

18 years of age or older. Their duties included the control of the apparatus at fires and restoring the equipment to good order afterward.

The town of Portsmouth was incorporated as a city in 1850, and in 1852, City Ordinance No. 21 established the duties and powers of the board of engineers, chief engineer, assistant engineers, and clerk. It also addressed the organization and duty of the city's fire companies and the policies governing its members and apparatus. Even as the community prospered in the midst of the Industrial Revolution, all fire personnel were on-call volunteers. By 1858, the city would boast a complement of six fire engines known as hand tubs. The engine companies adopted names such as Governor Langdon, Piscataqua, Atlantic, and Granite State. In 1863, the department would purchase its first hook and ladder carriage, necessitating the construction of a new firehouse on Court Street. A year later, the fire department would take delivery of its first steam fire engine, which was promptly named the Sagamore. It was soon followed by the Dearborn and Kearsarge. The Kearsarge gained special notoriety when in 1872 the engine and its fire company, along with the crew of the Dearborn, responded to the great Boston fire of 1872 and saved the Old South Church. By 1880, the Moses H. Goodrich and Colonel Sise steamers were also in service.

The department would hire its first permanent member, engineer Eugene Hoyt, in 1891. He was the driver of chemical engine No. 5. However, it would not be until 1927 that the department hired its first permanent chief engineer (fire chief). In 1900, the department would consist of 6 permanent firefighters and 93 call men. Chief Engineer John D. Randall also reported that the new engine house on Bartlett Street "ought to be finished as soon as the finances will permit." The first gasoline-powered fire engine to serve the city was purchased in 1912. By 1921, the department would be fully motorized, and all apparatus, except for the steam fire engine Sagamore, would be housed in a single central fire station that was completed a year earlier. Innovations and opportunities influenced the direction of the department following World War II and continued through the 1960s. In 1944, the department purchased an ambulance and began providing emergency medical response the following year. The department was also introduced to self-contained breathing apparatus (SCBA) during the later part of the 1950s. Throughout the 1950s and 1960s, the city would experience significant growth as both the Portsmouth Naval Shipyard and the newly opened Pease Air Force Base attracted military personnel and civilian workers to the community. Improved fire protection was required of the many residential neighborhoods that popped up throughout the city; hence in 1967, a new sub–fire station was completed and staffed on Lafayette Road.

The 1970s were marked by frequent and major fires. As the decade came to a close, the fire dollar loss would hit a record high of $874,000. That record was obliterated during the 1980 fiscal year when the largest fire loss in recent history occurred. Claims for that year exceeded $2 million. Some 21 structure fires impacted the dollar loss, with the largest single loss attributed to a blaze on October 13, 1980, that destroyed the McIntosh Block. The city council authorized the department to increase the staffing to a total of 60 full-time personnel in 1986, and the following year the establishment of the Portsmouth Fire Commission was approved by city vote. The commission was charged to work on policy and administrative matters within the department, being a level between fire chief and city council. In 1993, the department began covering the city with an additional fire station. The station on what is now International Drive was previously home to the air force's structural station on the former Pease Air Base.

Responding to over 5,000 calls a year, the Portsmouth Fire Department of today provides fire protection, advanced life support ambulance, fire prevention, hazardous materials response, maritime, and other special rescue services to the seacoast community of Portsmouth. The department provides these services through a complement of 61 full-time members. It is through the men and women of today's Portsmouth Fire Department, not unlike the perseverance of the night watchmen, the service of the early fire companies, or the resolve of those who fought some of the city's largest fires, that pride is constantly demonstrated and traditions forever preserved.

One

Early Fire Companies and Engine Houses

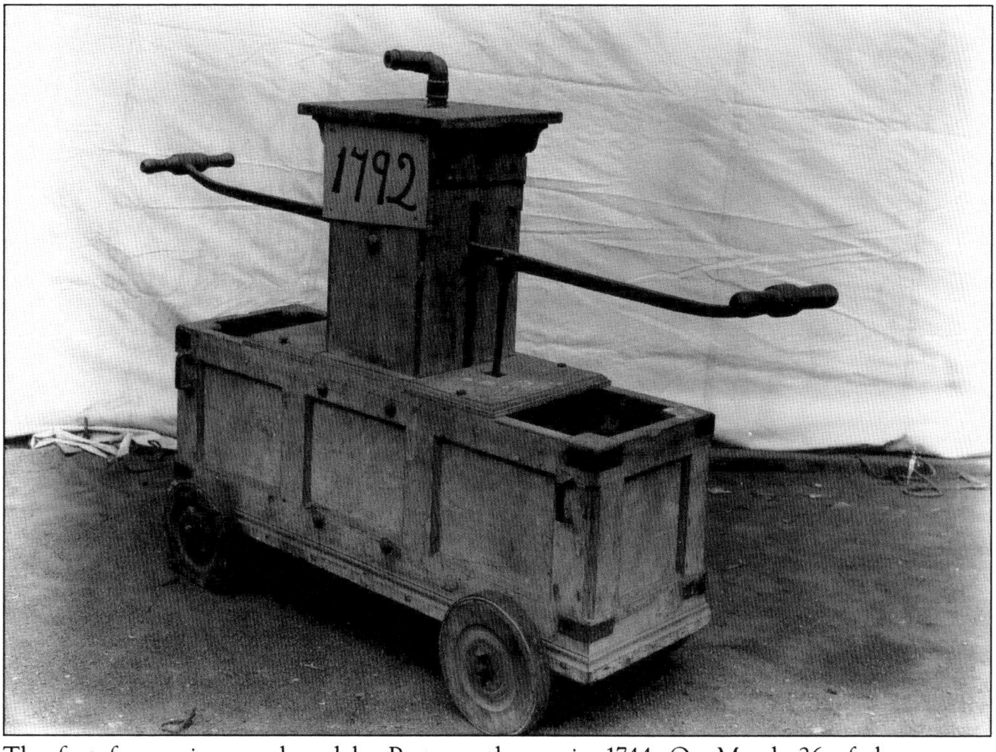

The first fire engine purchased by Portsmouth was in 1744. On March 26 of that year, a committee was established to purchase a "New Ingine." This hand engine, which was owned by the Portsmouth's Veteran Firemen's Association and did service in Philadelphia, is pictured in front of the Central Station on Court Street around 1920. (Courtesy of Strawbery Banke.)

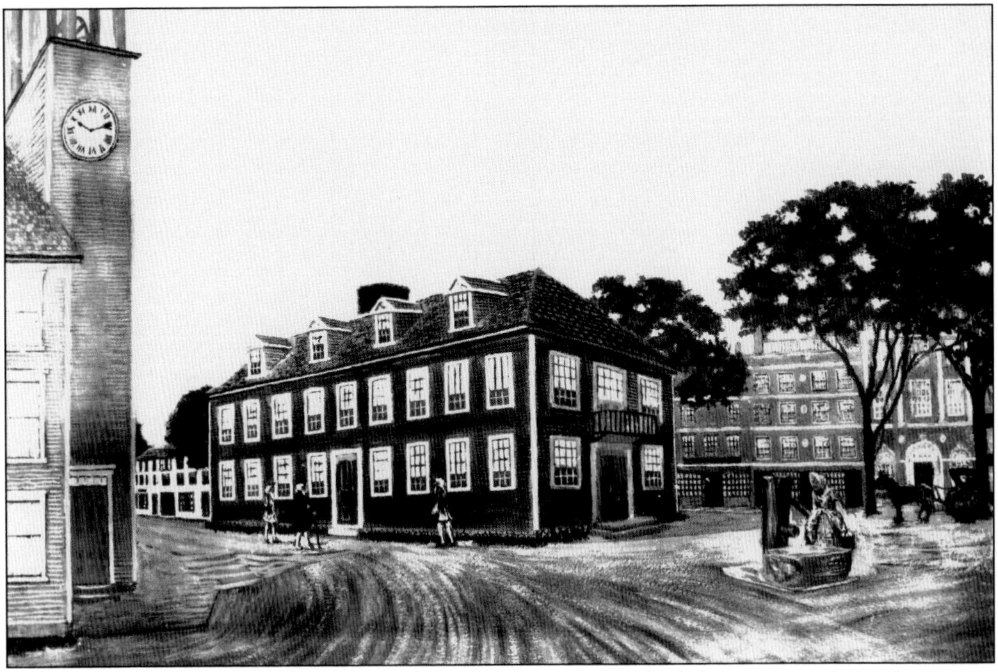

In 1758, the provincial legislature passed an act for organizing firefighting throughout the colony of New Hampshire. Towns were authorized to form fire companies and elect officials to oversee these companies. That same year the fire companies of Portsmouth had a room in the recently constructed statehouse located in what is now Market Square. (Courtesy of Strawbery Banke.)

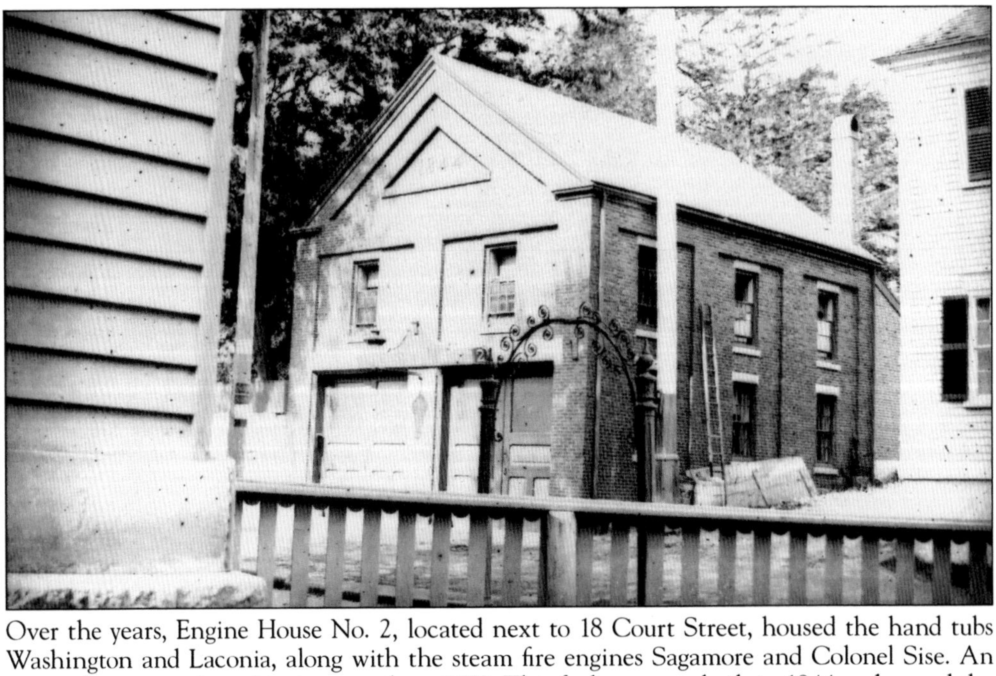

Over the years, Engine House No. 2, located next to 18 Court Street, housed the hand tubs Washington and Laconia, along with the steam fire engines Sagamore and Colonel Sise. An engine house stood on this site as early as 1802. This firehouse was built in 1844 and served the city through the end of the 19th century. (Courtesy of Strawbery Banke.)

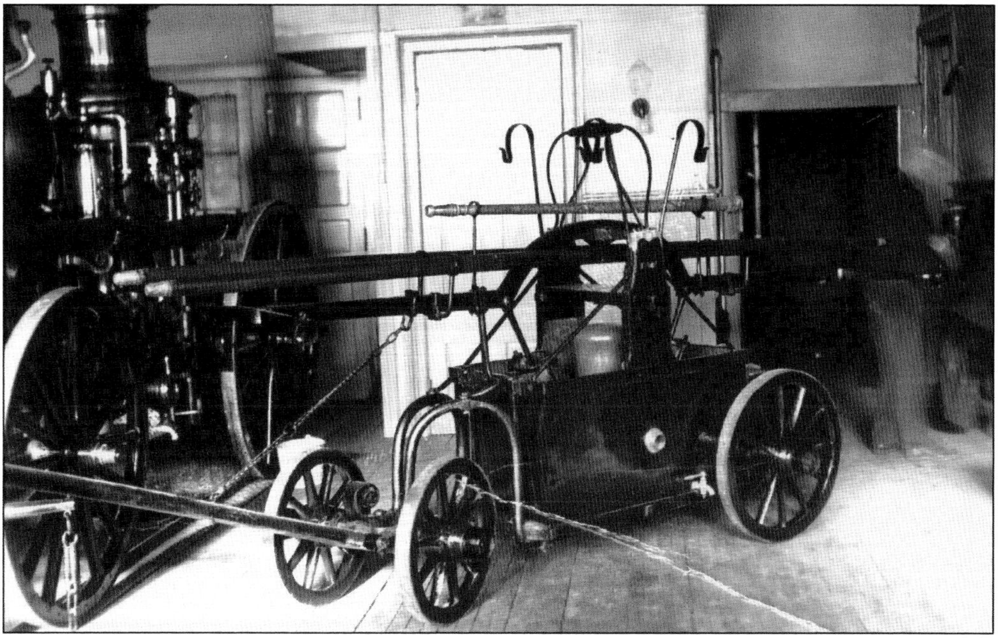

According to Col. William H. Sise, ex-mayor, fireman, and engineer of Portsmouth, "Washington No. 2 was Portsmouth's first suction engine. The men not knowing what the suction was for, it lay about the house useless, and at fires they filled the engine from water buckets. One day some one accidentally found out what the suction was intended to do and thereafter the firemen devoted to its proper care." This is a photograph of one of Portsmouth's hand-drawn suction engines, taken sometime around 1865.

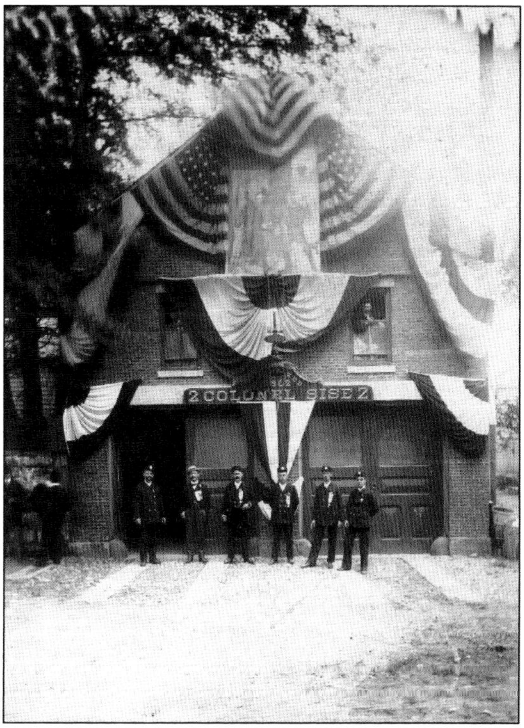

Portsmouth firehouses were adorned with bunting and flags during special occasions such as firemen musters, parades, and other civic celebrations. In this undated image of the Colonel Sise steam fire engine house, a large banner framed by American flag bunting can be seen depicting a vivid illustration of Portsmouth's bravest battling the "fire fiend." (Courtesy of the Connors family.)

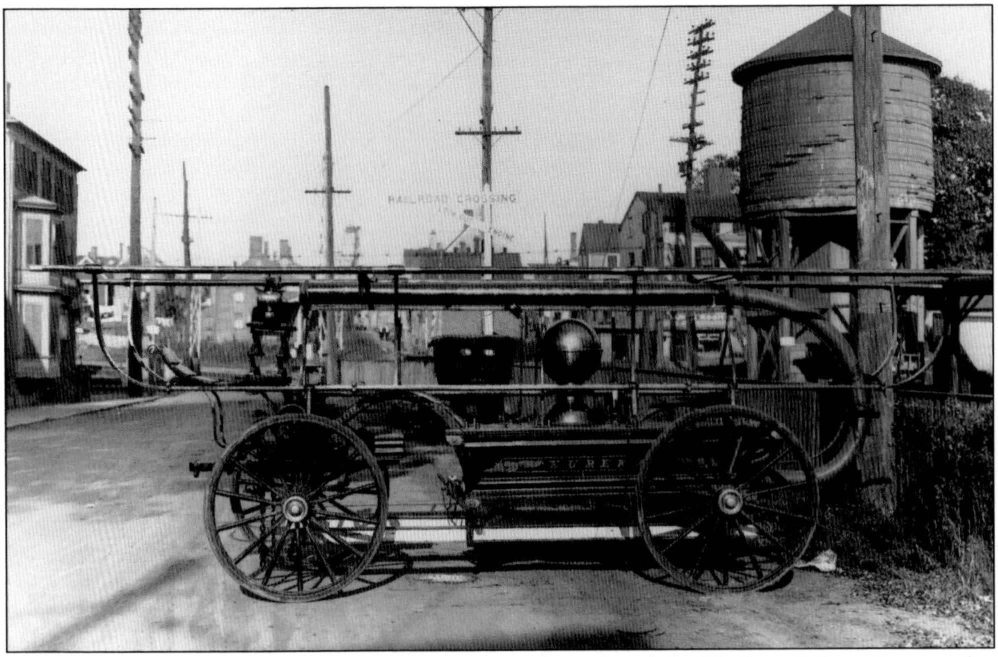

Eureka No. 1 was a hand tub that was built in 1872. It served the Hudson, Massachusetts, fire department until 1907, when it was sold to the Franklin Pierce Veterans Firemen's Association of Portsmouth. Pictured here in Portsmouth near the railroad tracks, the Eureka never served as a Portsmouth Fire Department engine but was an outstanding muster piece. (Courtesy of Strawbery Banke.)

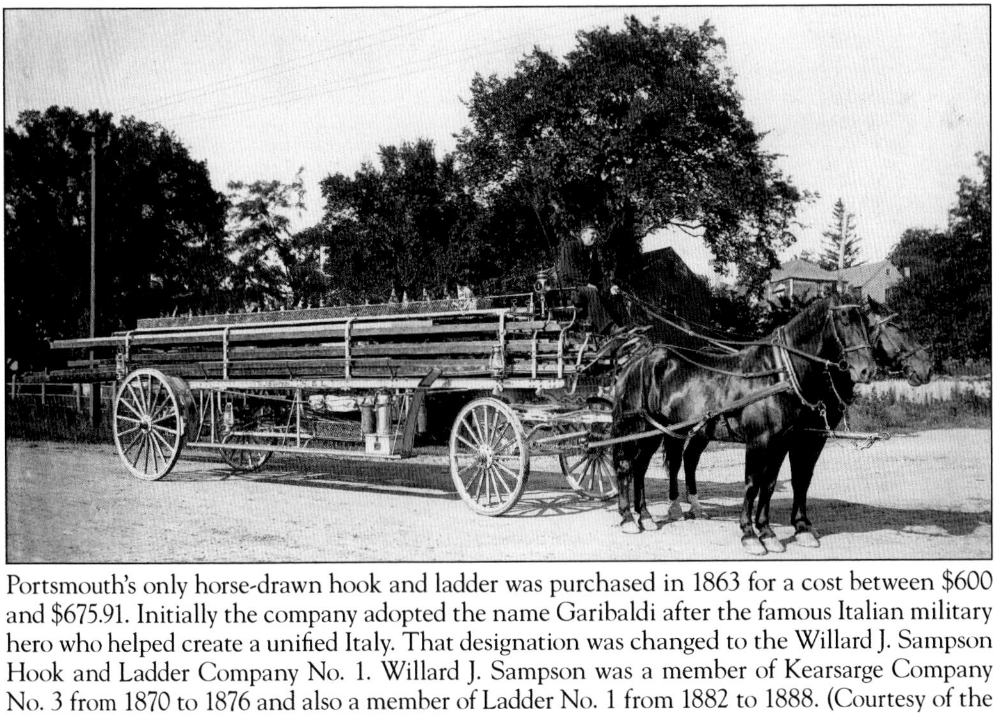

Portsmouth's only horse-drawn hook and ladder was purchased in 1863 for a cost between $600 and $675.91. Initially the company adopted the name Garibaldi after the famous Italian military hero who helped create a unified Italy. That designation was changed to the Willard J. Sampson Hook and Ladder Company No. 1. Willard J. Sampson was a member of Kearsarge Company No. 3 from 1870 to 1876 and also a member of Ladder No. 1 from 1882 to 1888. (Courtesy of the Connors family.)

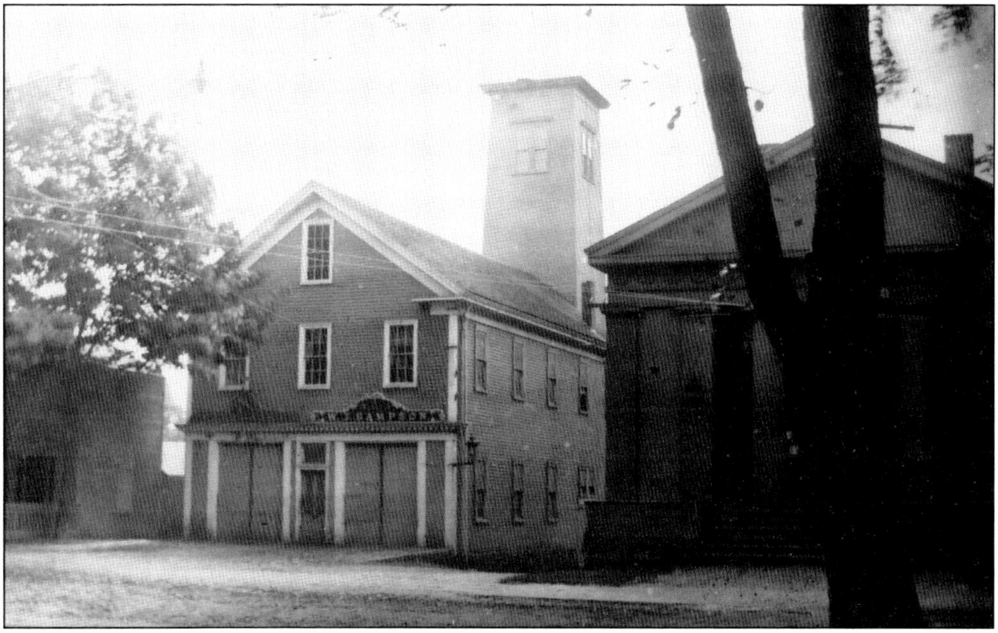

Constructed in 1863 at a cost of $3,558.81, the Central Engine House was built on a lot just east of the courthouse specifically for the use of the new hook and ladder carriage and engine companies. The services provided by the department in 1863, as reported by Chief Engineer John Moran, consisted of three fires and one false call. He also added that "The Engine houses and the Hook & Ladder house are all in good condition, convenient and comfortable." No longer a fire station, the building still stands today next to the Central Fire Station and is home to Baker and Wright Auto Electric Service. (Courtesy of Strawbery Banke.)

Portsmouth planned the Return of Sons homecoming event in 1853 for the city's native sons and daughters. While many came from Boston and New York, some journeyed from as far away as Florida and Wisconsin. The Return of Sons was again celebrated in 1873 and 1883. In this photograph taken in July 1873, men from the Kearsarge and Dearborn Fire Companies pose in front of the Court Street station under a patriotically decorated evergreen arch.

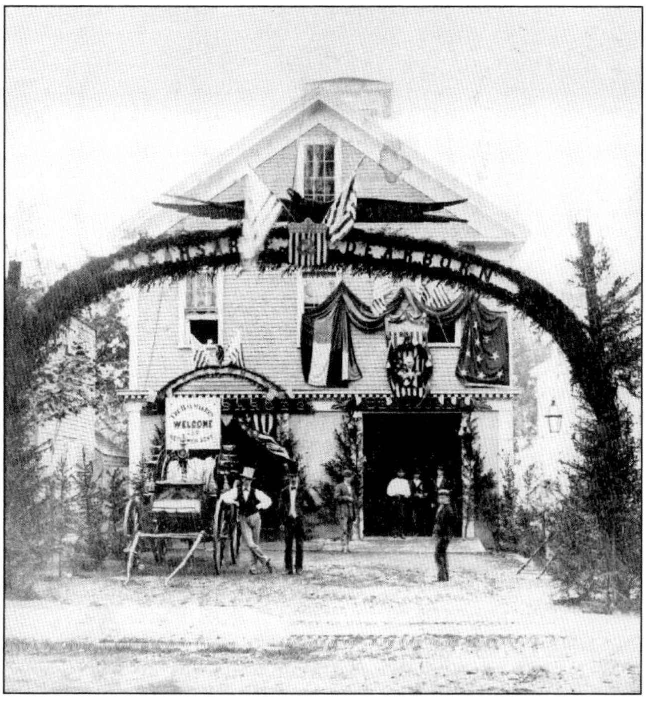

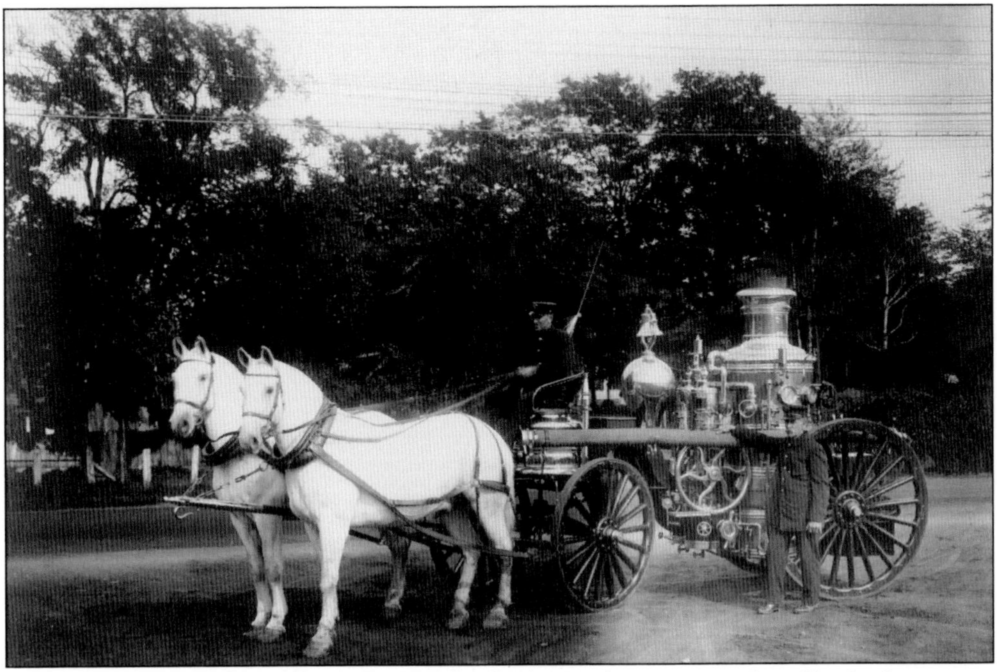

Kearsarge Steam Fire Engine (SFE) No. 3 was the third steam fire engine purchased by the department and was named for the USS *Kearsarge*, a sloop of war built and launched at the Portsmouth Naval Shipyard that gained notoriety for her defeat of the Confederate commerce raider CSS *Alabama* during the American Civil War. The engine responded out of the Court Street station next to the courthouse from 1870 through 1919. Pictured here is the Kearsarge with driver William T. Fernald.

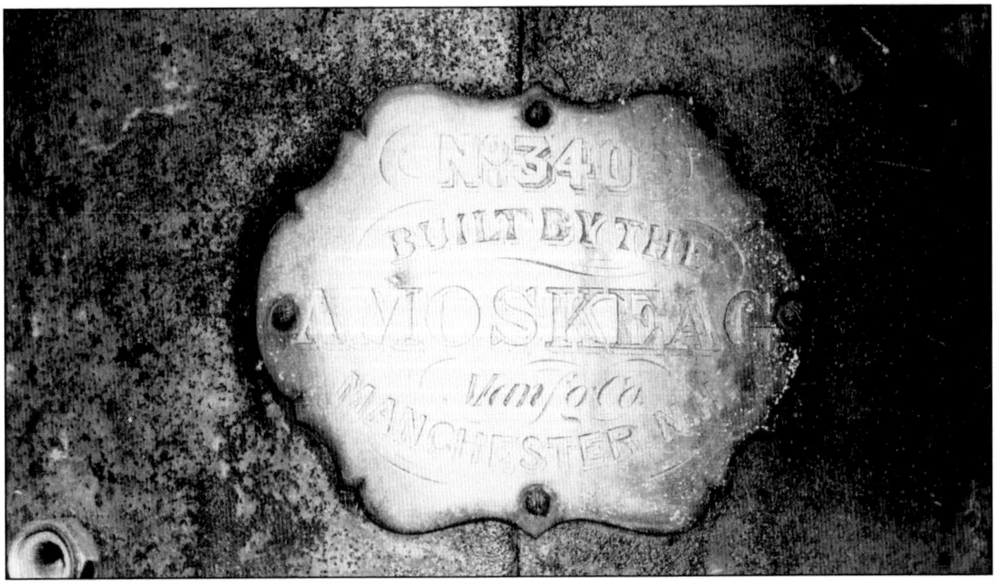

Portsmouth's Kearsarge SFE No. 3 was built by the Amoskeag Manufacturing Company of Manchester. The weight of the engine, when drawn to fires, was 6,400 pounds. By the next year, the Kearsarge SFE Company No. 3 would consist of 27 firefighters. Seen here is the original manufacturer's plate, identifying the steam engine as the 340th built by Amoskeag.

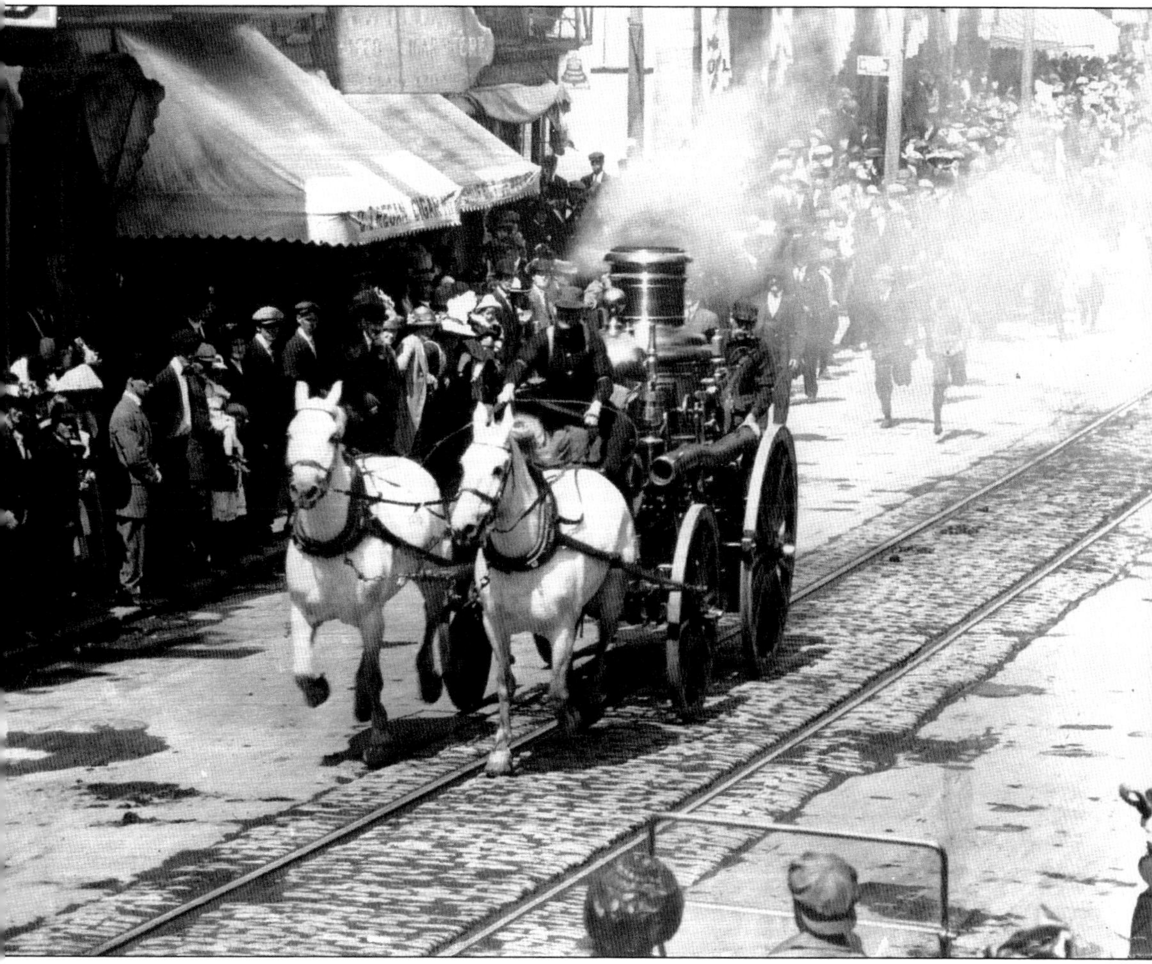

Engineer George N. Jones rides on the rear platform of Engine 3 as Duncan Cameron drives the horses and steam engine to a fire on Dover Street just as the Memorial Day parade was about to start in this May 30, 1913, photograph. The local newspaper praised Cameron for his ability to handle the horses through the enormous crowds that broke and ran from the parade route toward the fire when box 43 sounded without injuring a single person. The fire at 6 Dover Street gutted the top floor of a two-and-a-half-story wood frame home shortly after 2:00 p.m. Firefighters arrived at the address to find the entire roof ablaze, and at one time, it looked as if the sparks and embers carried by the high wind would set the surrounding structures aflame. Chief William F. Woods considered sounding another alarm, but the firemen made quick work, and at 2:40 p.m., the all-out signal was sounded.

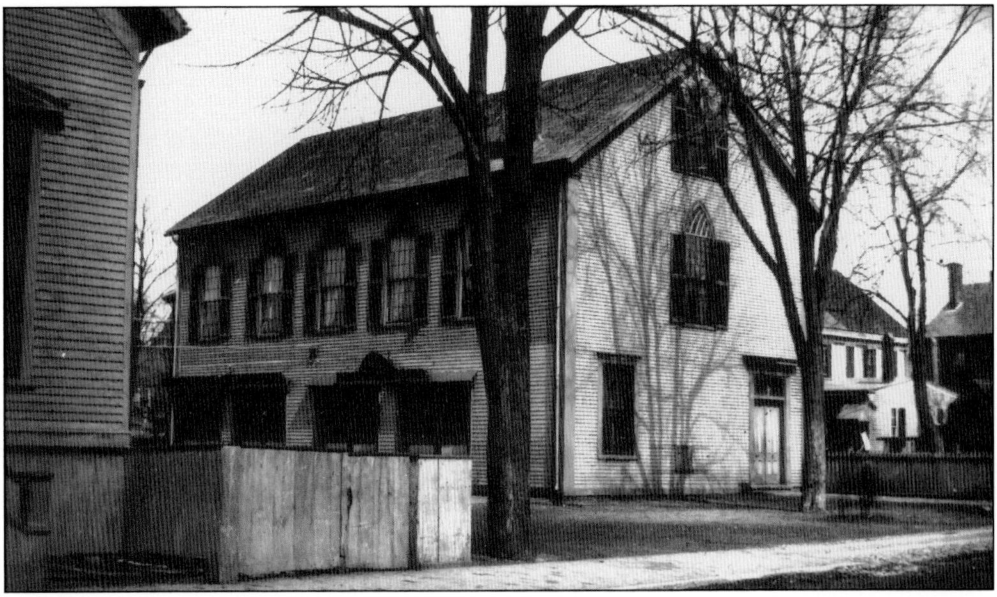

Situated next to the Advent Church and the Hanover Street Grammar School, the Hanover Street fire station was built in 1876. Located at 42 Hanover Street, the station was initially home to the Dearborn SFE Company. Over the years, Moses H. Goodrich No. 4, Eagle No. 5, and Sagamore No. 2 would run out of this engine house. (Courtesy of Strawbery Banke.)

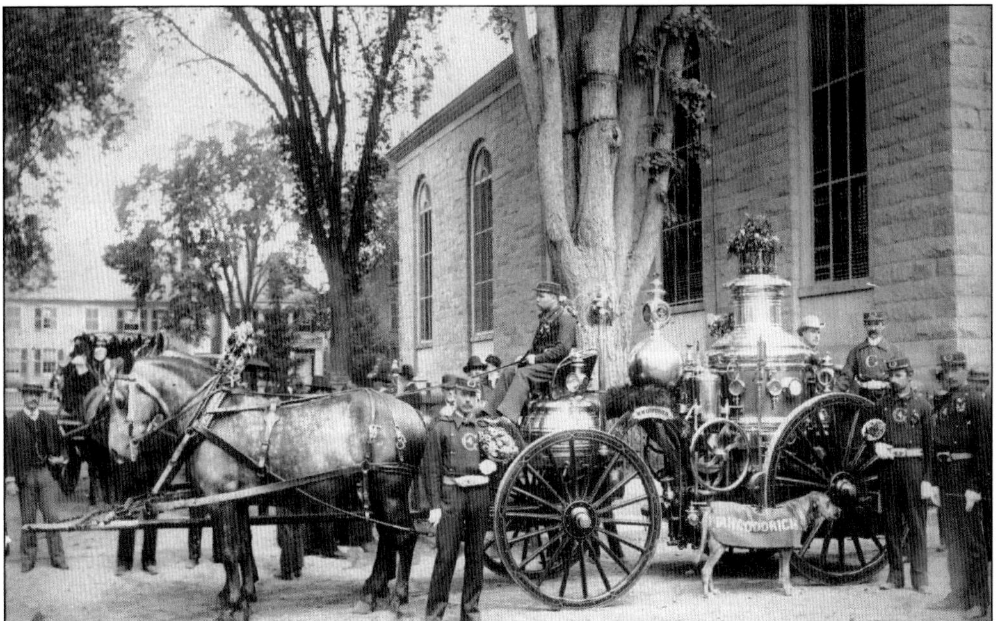

The steam fire engine Moses H. Goodrich was delivered to the Portsmouth Fire Department in July 1876. Manufactured by the S. S. Nichols and Company of Burlington, Vermont, the engine was stationed at the Hanover Street firehouse until it closed in 1920. The fire company's namesake, Moses H. Goodrich, was the mayor of Portsmouth in 1874 and 1875. He was also a member of New Hampshire State Senate 24th District from 1885 to 1886.

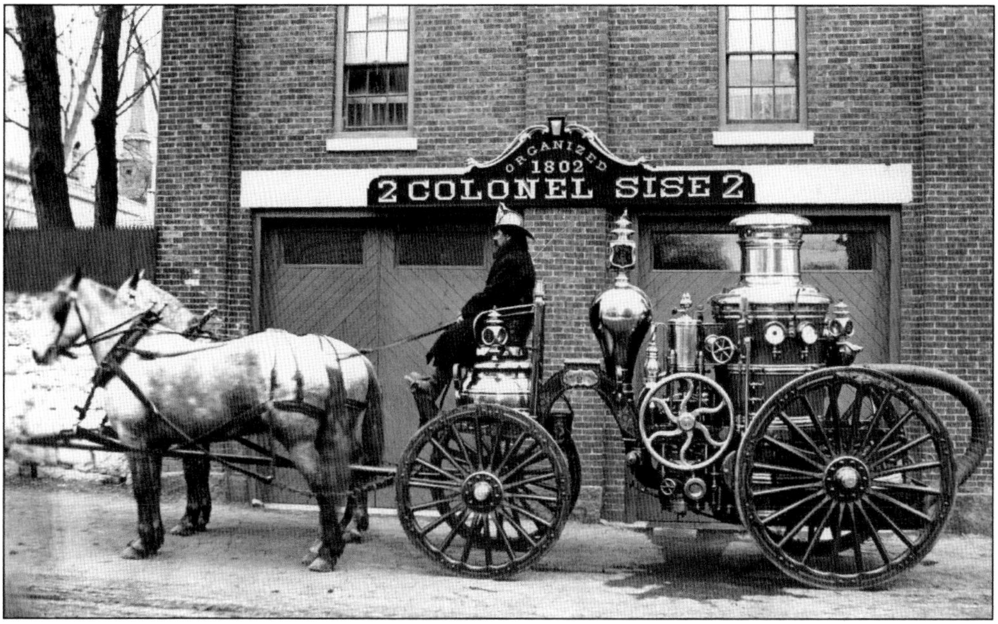

The fifth of the Portsmouth steamers, the Colonel Sise, was built by the Amoskeag Manufacturing Company of Manchester and placed in service in 1880 at the engine house on lower Court Street. "We'll Try" was the motto of the company named for Col. William H. Sise, fireman, engineer, and mayor of Portsmouth from 1878 to 1881.

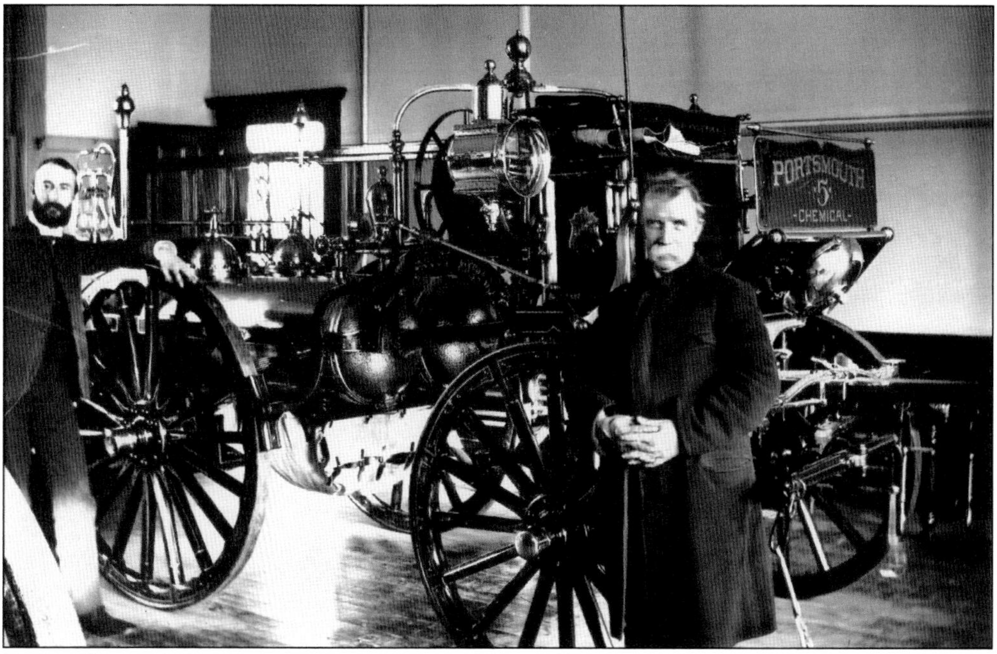

Eagle Chemical Engine No. 5, manufactured by Holloway of Baltimore, was delivered to the department in 1890. The apparatus contained chemical tanks, which used acid combined with soda dissolved in water to start a chemical reaction that would produce carbon dioxide. In this process, the carbon dioxide expanded, pressurizing the tank and propelling the entire mixture out of the hose line and onto the fire.

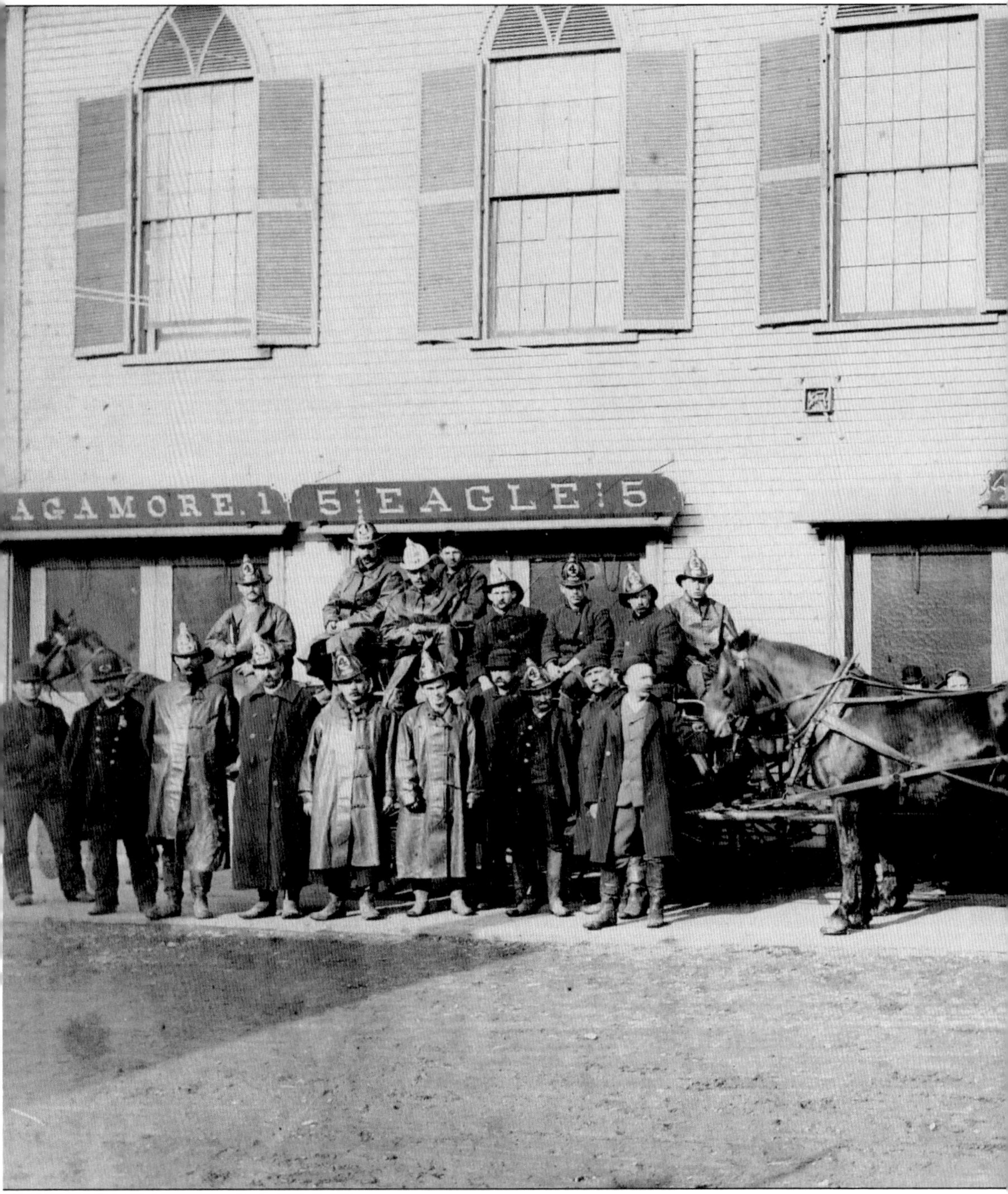

During 1893, the Portsmouth Fire Department responded to 18 alarms, 2 of which were stills by the Chemical, and one to the town of Exeter where "assistance was rendered." On March 14, 1893, box 58 sounded, sending the Moses H. Goodrich SFE and Hose Wagon Company No. 4 to Exeter. Pictured here is the company on its return from the fire. Also housed at the Hanover Street station at this time were both Eagle Chemical No. 5 and in reserve, Sagamore SFE No 1. It is worth noting that during the 19th century and early part of the 20th century assisting another

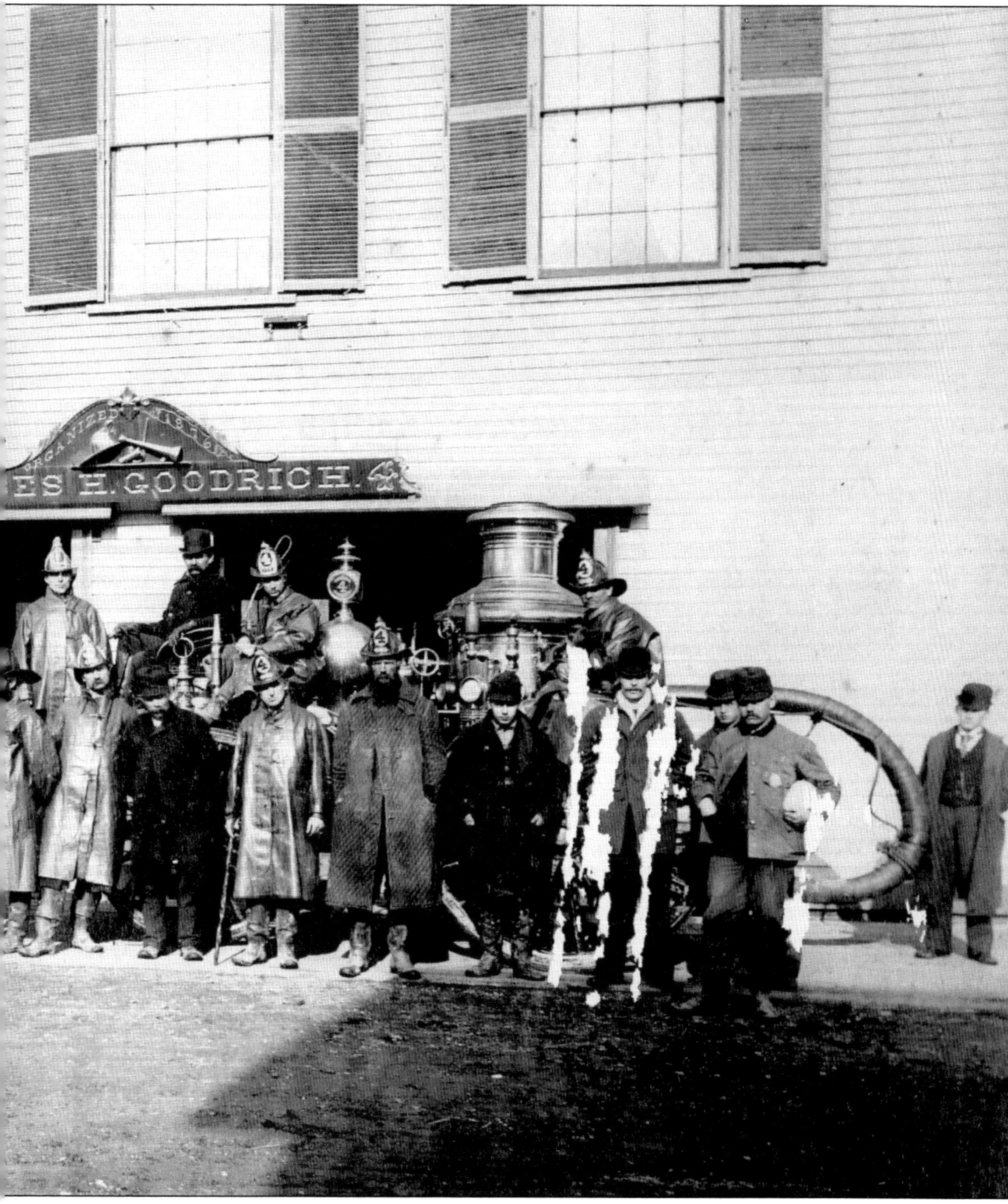

fire department was rare because of the disastrous results that could occur if a city or town did not maintain its own appropriate level of fire protection. Therefore, Portsmouth City Ordinance No. 21 contained the provision that "No part of the apparatus of the Fire Department shall be carried out of the city without the permission of the Chief Engineer, or in his absence, of the Mayor; and the same shall be for no other use than the extinguishment of fires."

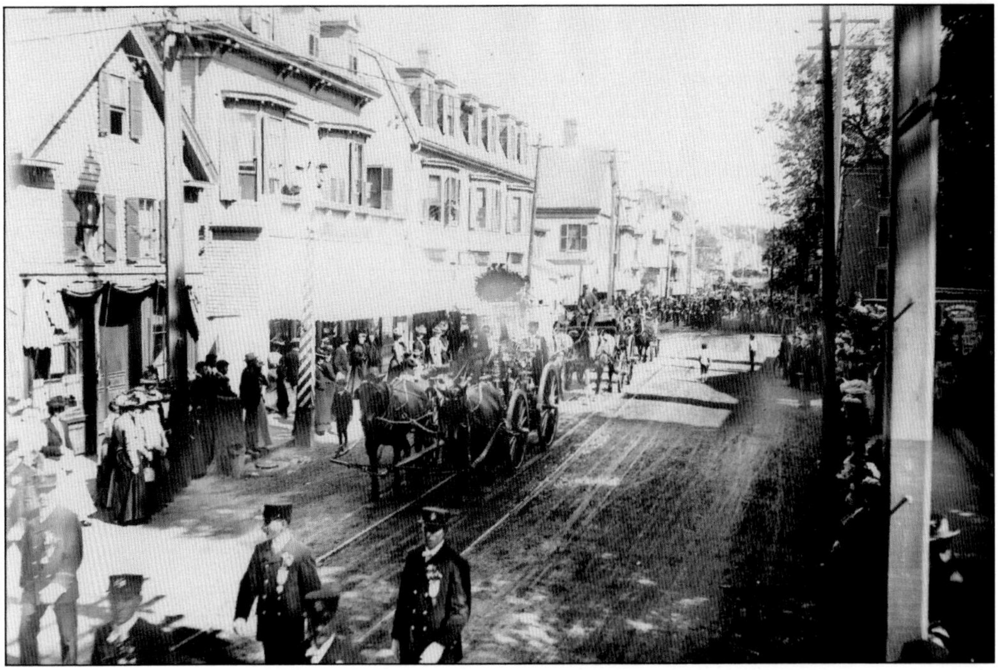

"In compliance with Section 3, Chapter 13 of the City Ordinances, the first annual parade of the department occurred on September 15, 1887 and I have no doubt that the members of the City Councils, as well as the citizens, generally believe it was a wise expenditure, bringing the department as it does into more social relations with our citizens," said Herbert Marden, chief engineer. These two parade images of the department's steam fire engines were taken on Thursday, September 26, 1901. (Courtesy of Strawbery Banke.)

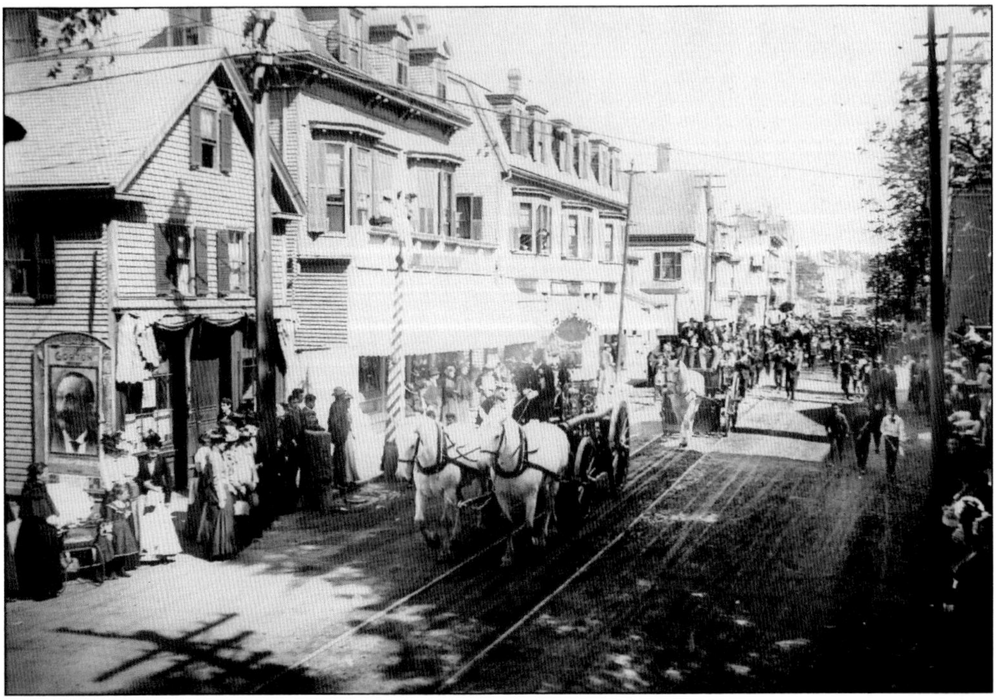

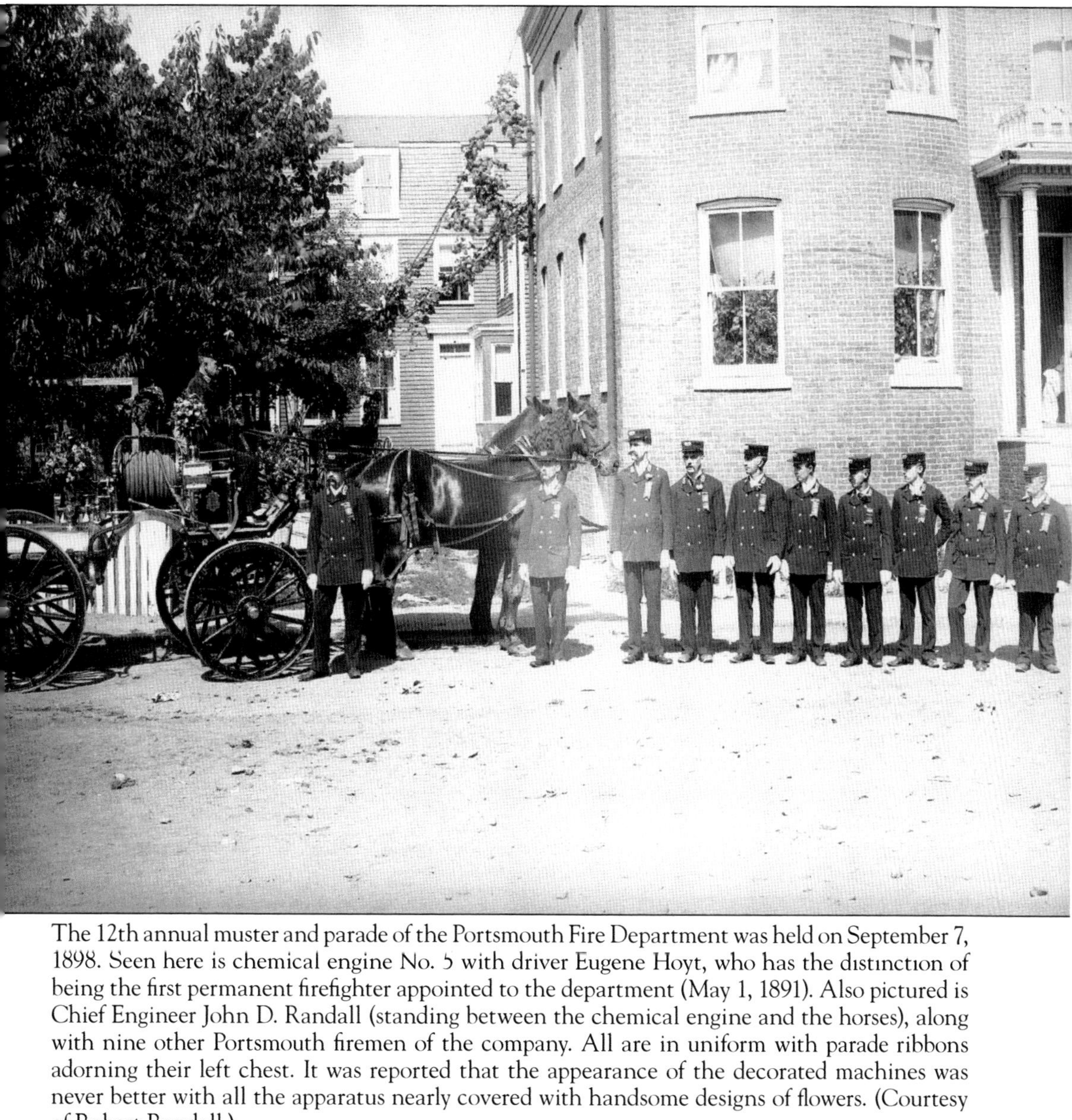

The 12th annual muster and parade of the Portsmouth Fire Department was held on September 7, 1898. Seen here is chemical engine No. 5 with driver Eugene Hoyt, who has the distinction of being the first permanent firefighter appointed to the department (May 1, 1891). Also pictured is Chief Engineer John D. Randall (standing between the chemical engine and the horses), along with nine other Portsmouth firemen of the company. All are in uniform with parade ribbons adorning their left chest. It was reported that the appearance of the decorated machines was never better with all the apparatus nearly covered with handsome designs of flowers. (Courtesy of Robert Randall.)

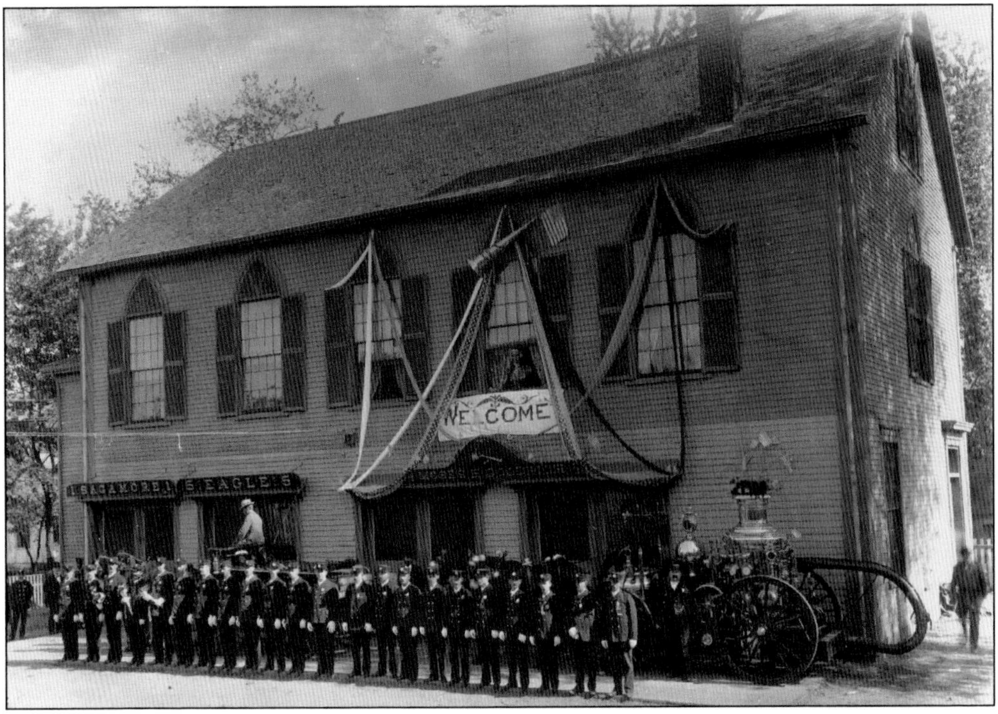

Portsmouth firehouses were the center of attention on Thursday, September 26, 1901, when Portsmouth celebrated Fireman's Day. At the stroke of 11:00 a.m., the fire alarm gave one bell, and a great parade of glittering machines was started. It was reported that never had the men or the engines looked finer. The straight, sturdy, and keen-faced firefighters in their natty uniforms of dark blue, their jaunty caps, and bright-hued badges made a grand display. The following day the fourth annual meeting of the New Hampshire Firemen's State Relief Convention was held in the city at Philbrick Hall, with a large representation from the various cities and towns of the state. Pictured here are the Moses H. Goodrich SFE Company at the Hanover Street station (seen above) and the Colonel Sise SFE Company at the lower Court Street engine house (seen below). (Courtesy of Strawbery Banke.)

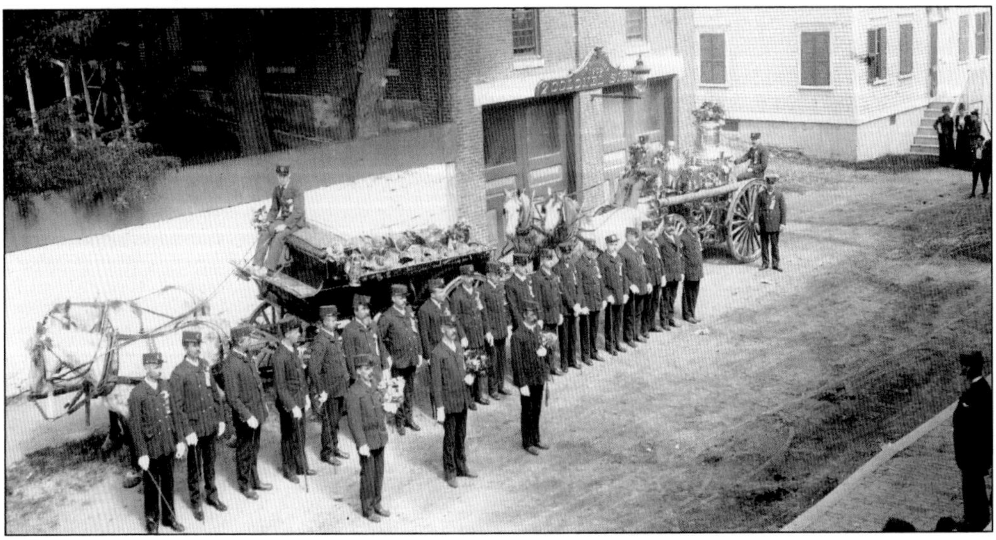

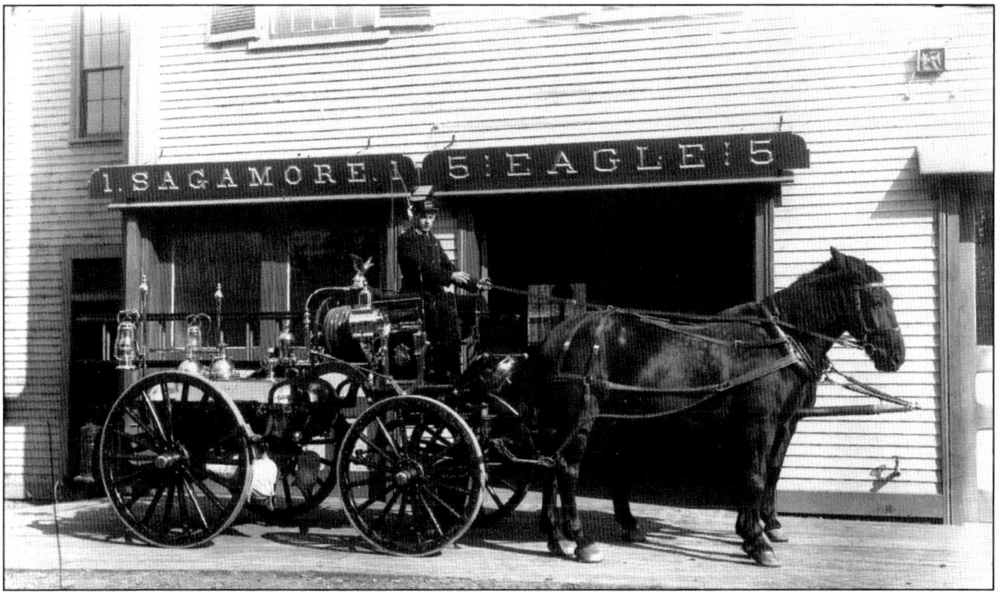

A pair of bay horses was purchased by a committee of the fire department for use on chemical engine No. 5 around 1896. These horses took the place of "a pair of blacks" that hauled the engine since it was purchased. Both horses were purchased in Boston for $250. In this photograph taken in 1905, Portsmouth's horse-drawn chemical engine No. 5 and its unidentified driver pose in front of the Hanover Street station. (Courtesy of Strawbery Banke.)

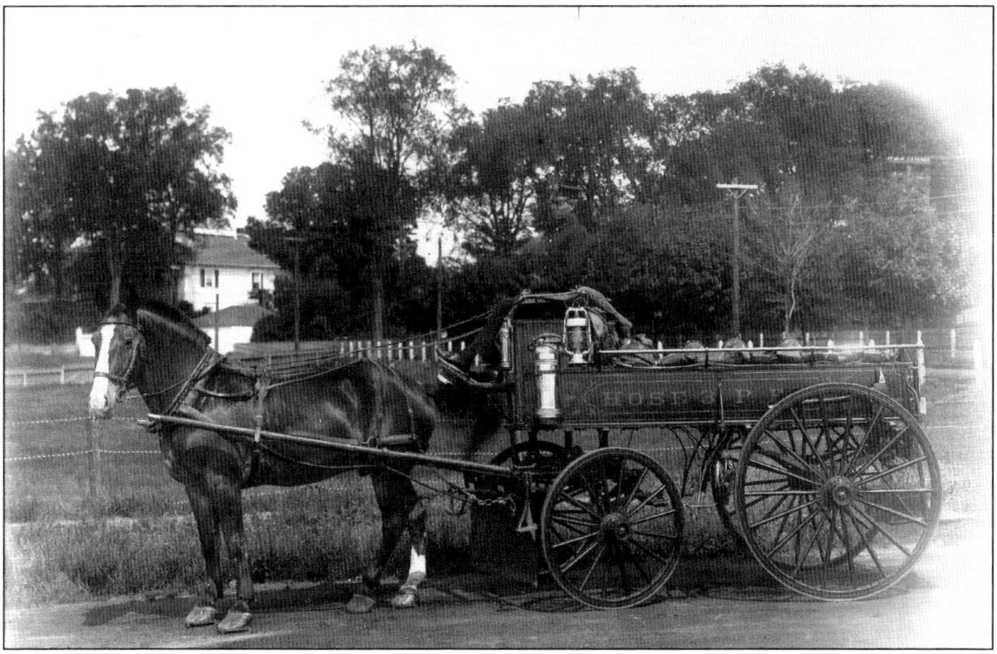

The schedule of property belonging to the fire department in 1900 included four hose wagons, one supply wagon, two hose reels, and two hose sleighs. The department also boasted 8,500 feet of serviceable two-and-a-half-inch cotton hose. Horse-drawn hose and supply wagons served a valuable role in not only transporting fire hose but also firefighting gear and equipment to fire scenes. (Courtesy of Strawbery Banke.)

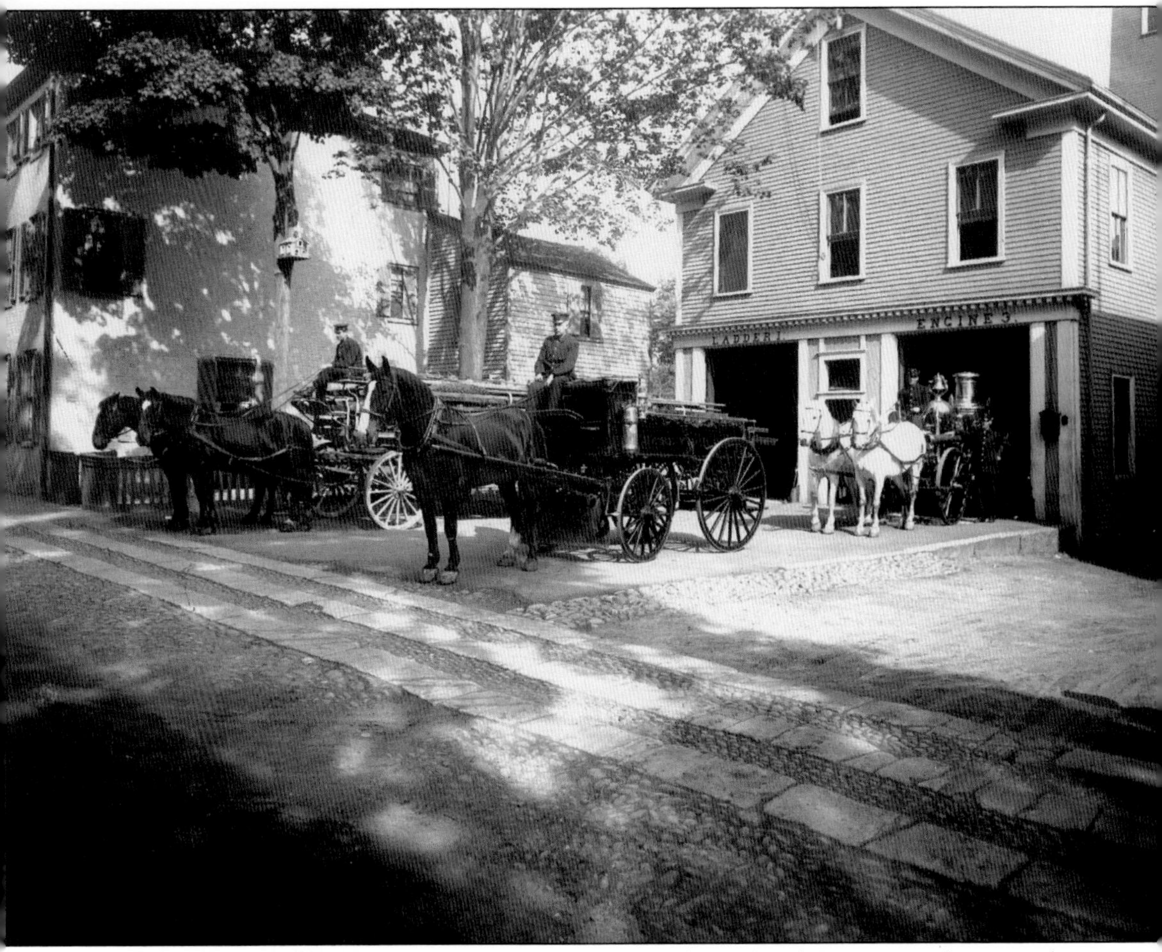

The engine houses of the department in 1900 were the Colonel Sise engine house on lower Court Street, the Hanover Street engine house, and the Central Engine House. Seen here in this image of the Central Engine House on Court Street are, from left to right, the Willard J. Sampson Hook and Ladder Company No. 1, hose wagon No. 3, and the Kearsarge SFE Company No. 3. In 1900, the department issued rules for exercising the horses used by the department. According to Chief Engineer John D. Randall, "It shall be the duty of the drivers of Engine No. 3, Hose No. 3, Chemical Engine and Hook and Ladder Truck to exercise the horses every day, weather permitting, except Sundays . . . In exercising, care must be taken to avoid colliding with other teams. In approaching corners, crossings, street car tracks, and in going down grade, the speed of the horses must be checked. Any driver violating these rules will be liable to suspension or discharge."

Two

THE AGE OF MOTORIZED FIRE ENGINES

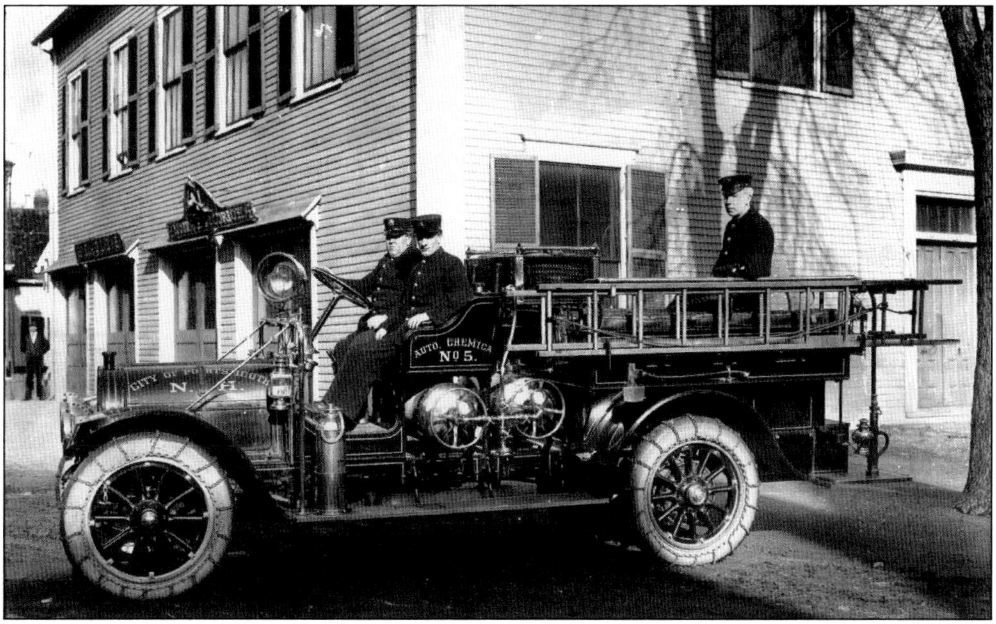

In 1912, the first gasoline-powered fire engine went into service in the city. The auto combination chemical and hose wagon was manufactured by the Pope-Hartford Manufacturing Company and boasted a four-cylinder engine capable of developing 50 horsepower. According to Chief Engineer John D. Randall, "The Auto-Chemical, put in commission in June, has fulfilled all that was claimed of it, both in upkeep and time saved in reaching the location of fires, especially long distances and the city has a valuable piece of fire-fighting apparatus in this machine." Pictured at the Hanover Street station are driver Francis Quinn, passenger George T. Cogan, and on the rear, Herman Crompton.

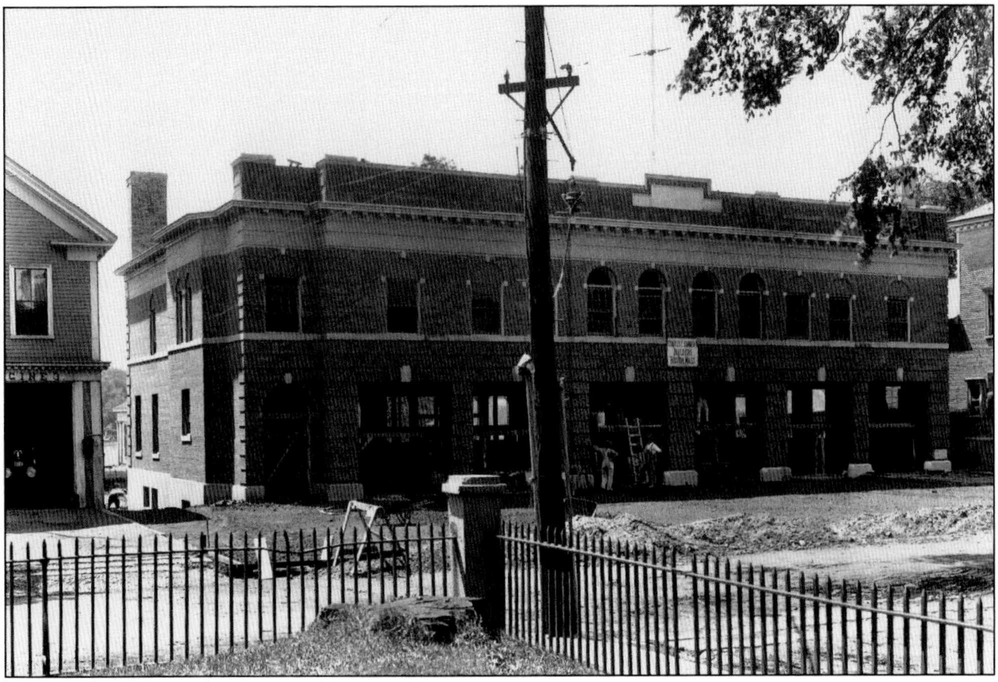

In 1912, Chief Engineer John D. Randall made the recommendation that a "modern and up-to-date engine house" be built to replace the Central Engine House on Court Street. In July 1919, the city council gave authorization to move forward with plans to build a new Central Fire Station for a cost of $85,000. By September 1919, it was determined that the location of the new station would be on the lot of the old courthouse that was just west of the current station on Court Street. The new station would be constructed of brick, have two stories, have six doors, and house all the apparatus except for the Sagamore Company. The upper floors would be used for the sleeping quarters for the permanent firemen, the company lounge room, and a clubroom for members of the various companies. The plans did not include any place for horses.

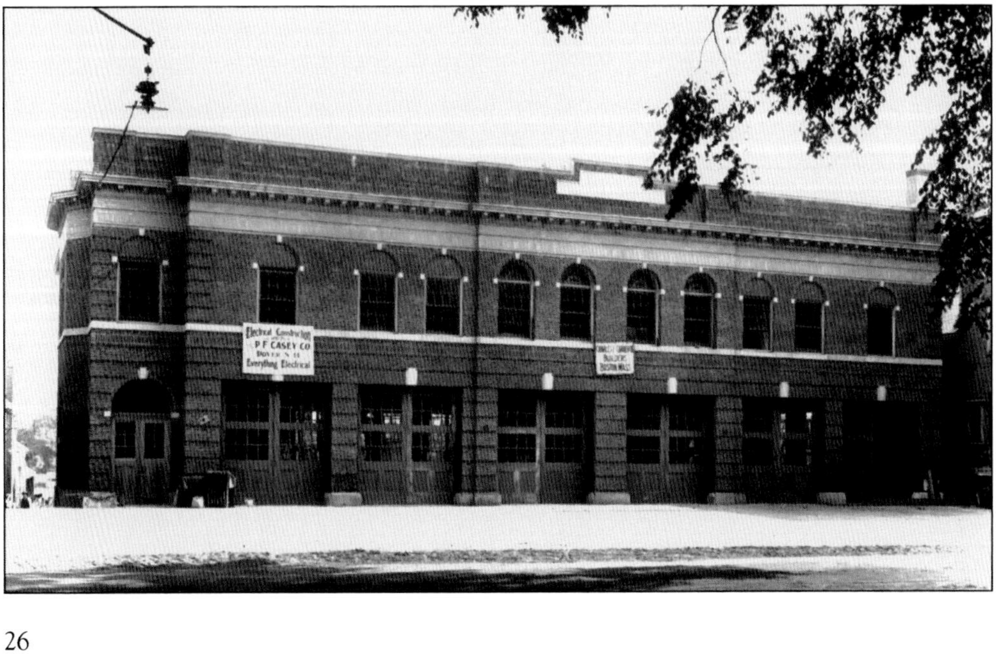

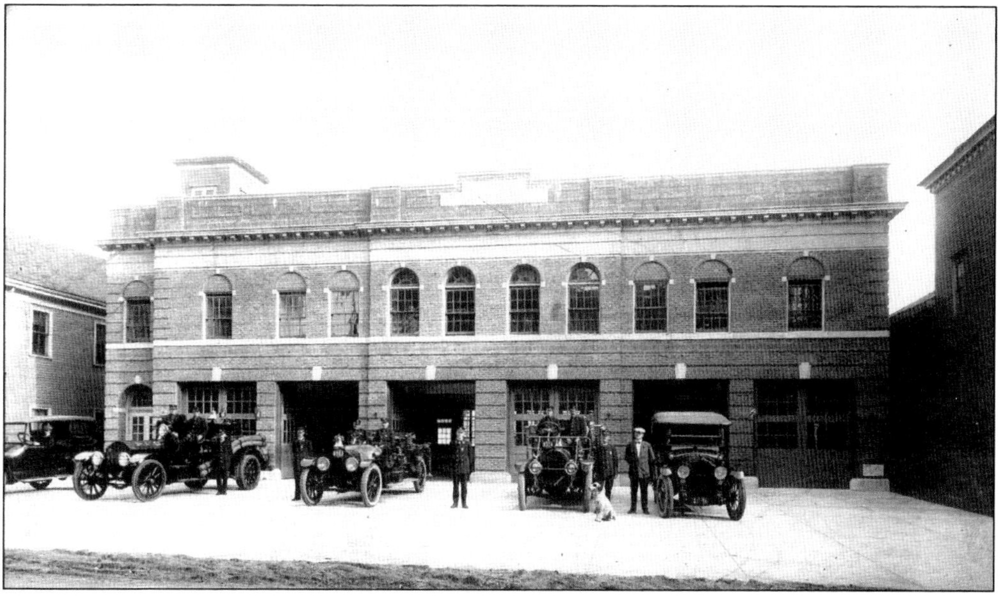

The apparatus of the Portsmouth Fire Department is on display in this image of the newly opened Central Fire Station taken sometime in 1920 or early 1921. In the photograph are, from left to right, a Buick touring car, the American LaFrance triple combination pumper, a converted Cadillac combination pumper, the Pope-Hartford auto combination chemical and hose wagon, and an unidentified touring car. (Courtesy of the Connors family.)

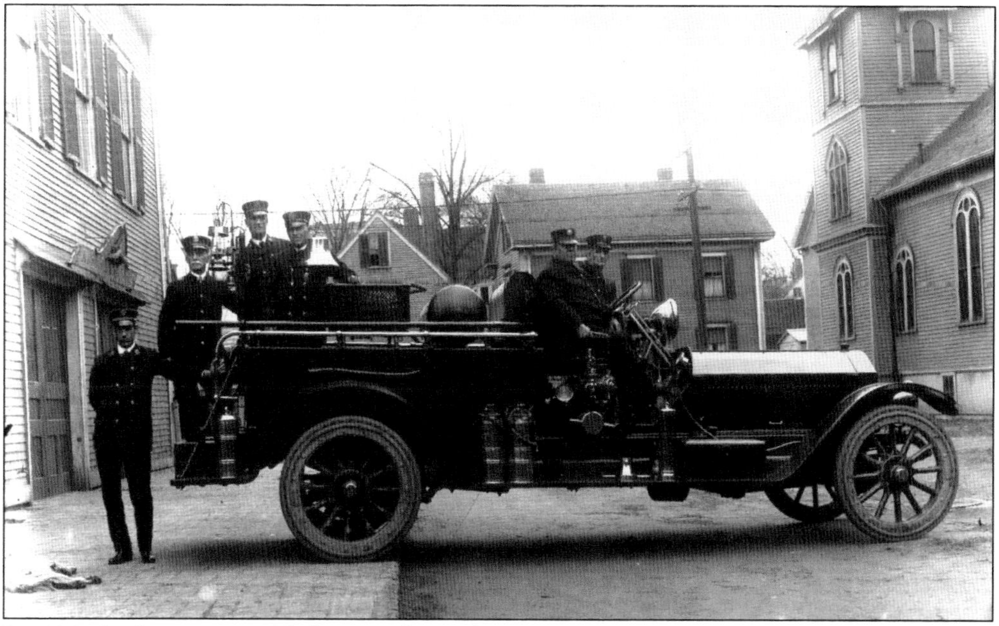

In 1920, the department added one triple combination, 750-gallon-per-minute pumper purchased from the L. H. Shattuck Company at a "very reasonable price." The American LaFrance pumper replaced one hose wagon and two steam fire engines and brought the complement of motorized apparatus to a total of three. This photograph of the engine and its company was taken in front of the Hanover Street station. To the right and across the street is the Advent Church. (Courtesy of Strawbery Banke.)

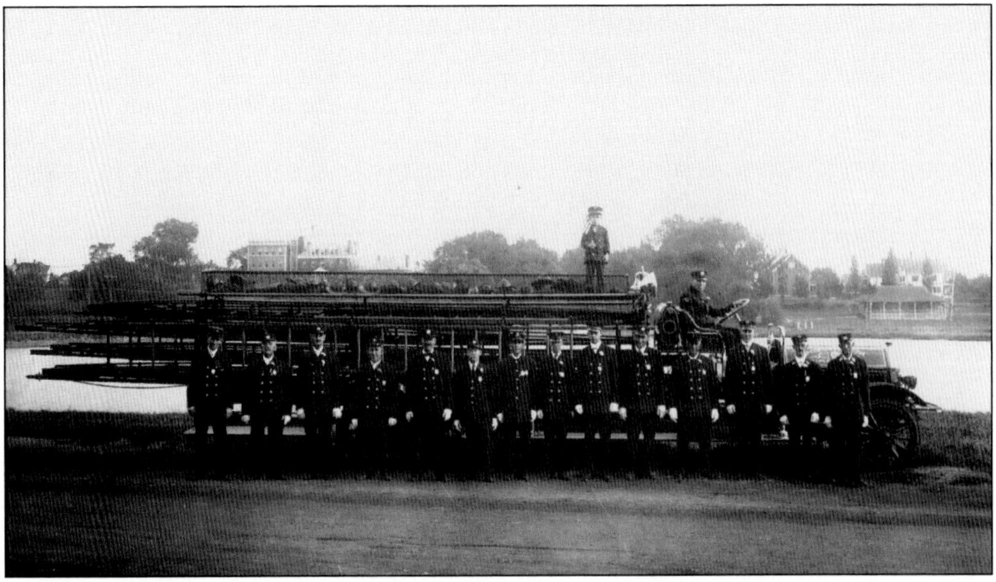

In service in 1921, this photograph of the Seagraves city service truck equipped with truss ladders was taken on September 28 of that same year near the shores of the South Mill Pond. In 1921, the department consisted of a chief, first assistant chief, second assistant chief, clerk, seven permanent men, and five call men assigned to Combination No. 5. The young mascot atop ladder No. 1 can be found in a number of photographs of this time period, but his name is a mystery.

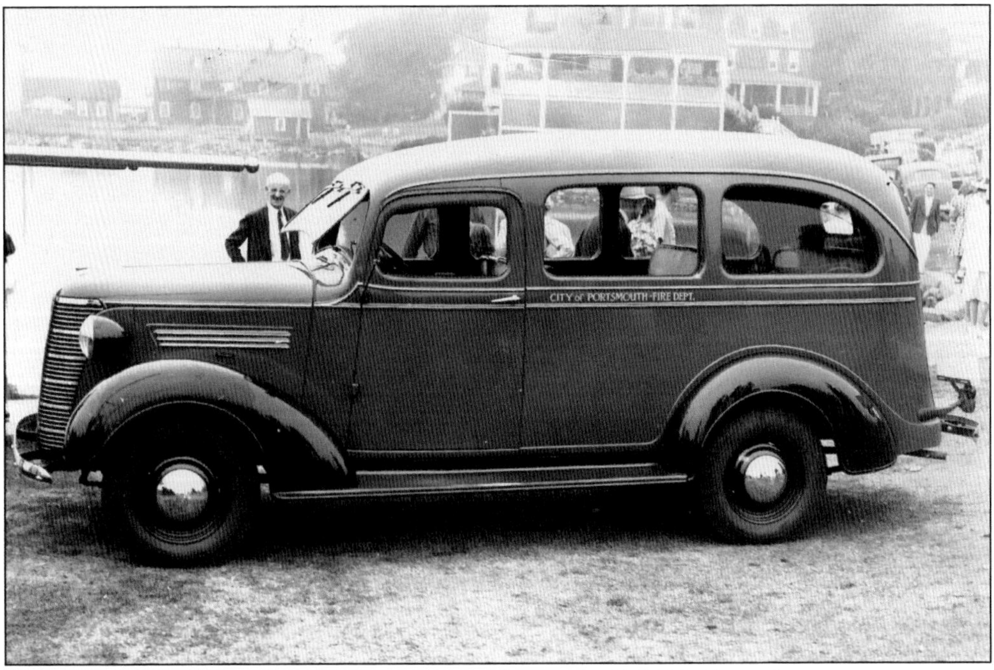

This photograph of Chief George T. Cogan and Portsmouth's 1938 Chevrolet station wagon was taken in 1939 at the Firemen's Field Day held at the Marshall House Grounds in York Harbor, Maine. The affair attracted fire departments throughout New England to this southern Maine coastal town. Events included a parade of fire apparatus and a firemen's muster.

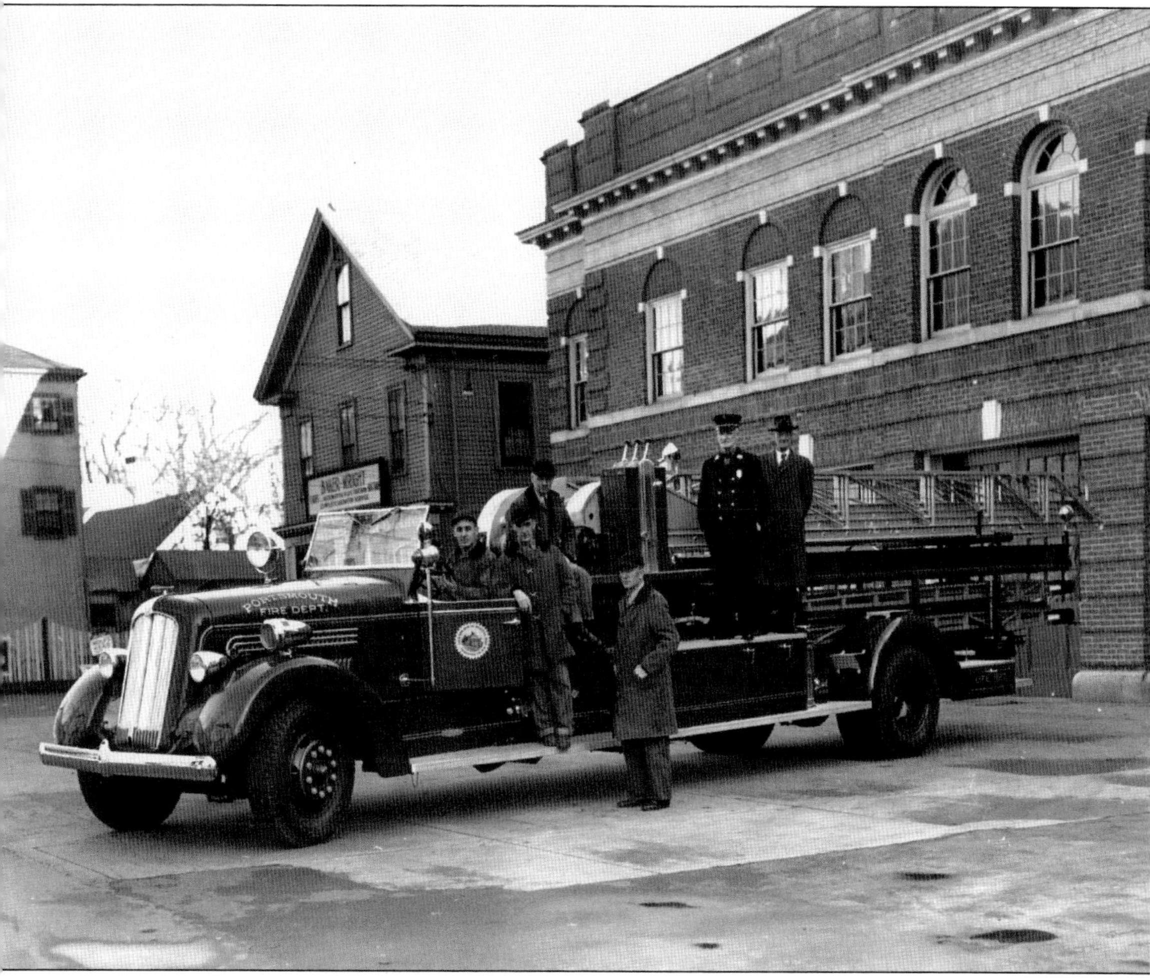

After a long campaign to secure funding from the city council, a new Seagraves 85-foot aerial ladder truck was delivered to the Portsmouth Fire Department on November 25, 1941. Within a week, the aerial truck was tested on the North Church in Market Square, and it was noted that when fully extended the ladder reached nearly nine feet above the clock. It was determined that the height would be sufficient throughout the city. Seen in this photograph of ladder No. 1 are, from left to right, driver John "Bill" Connors, firefighter Ernie Weeks, firefighter Fredrick R. "Ted" Crompton (seated), unidentified, Chief George T. Cogan, and unidentified. (Courtesy of the Connors family.)

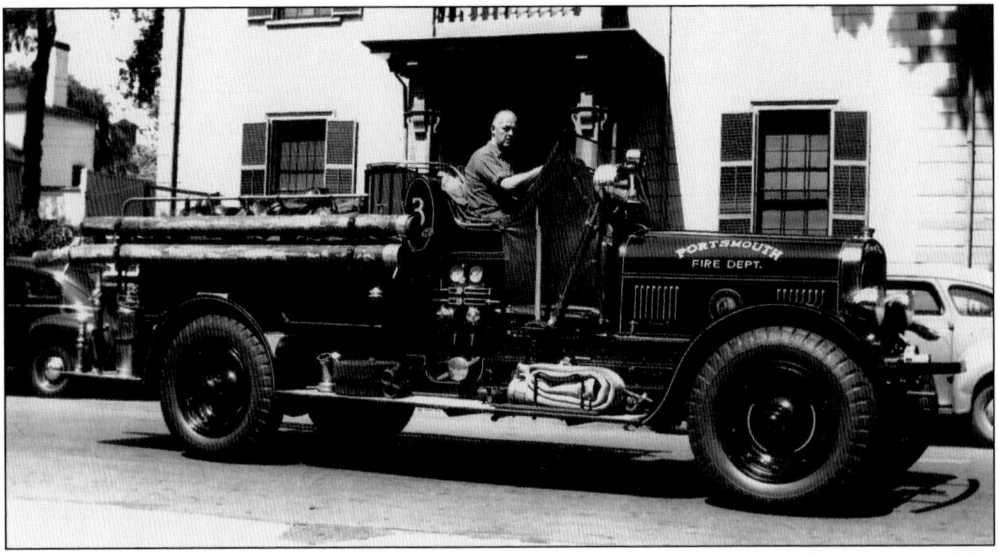

On August 31, 1944, Francis J. Carbone of Bradford, Massachusetts, captured six images of the Portsmouth Fire Department apparatus fleet. Pictured here are two of the six photographs: engine No. 3 with firefighter Frank E. Amazeen (seen above) and ladder No. 1 driven by firefighter Carl B. Akerley (seen below). Firefighter Frank E. Amazeen became a call man in 1924 and a permanent firefighter on August 6, 1941. Firefighter Carl B. Akerley was appointed as a permanent firefighter on September 28, 1942, and resigned to take a position with the New Hampshire Fish and Game Department in 1954. (Courtesy of Robert Randall.)

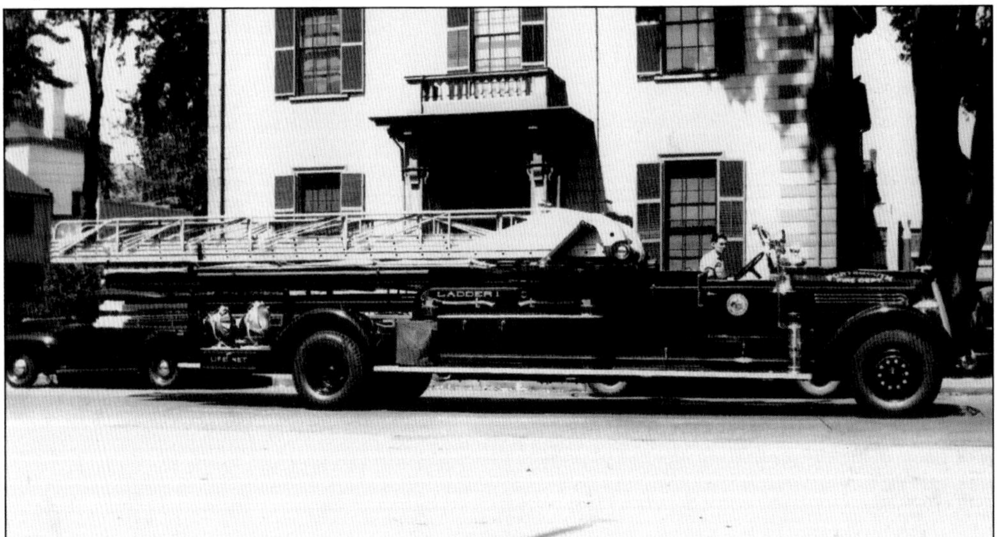

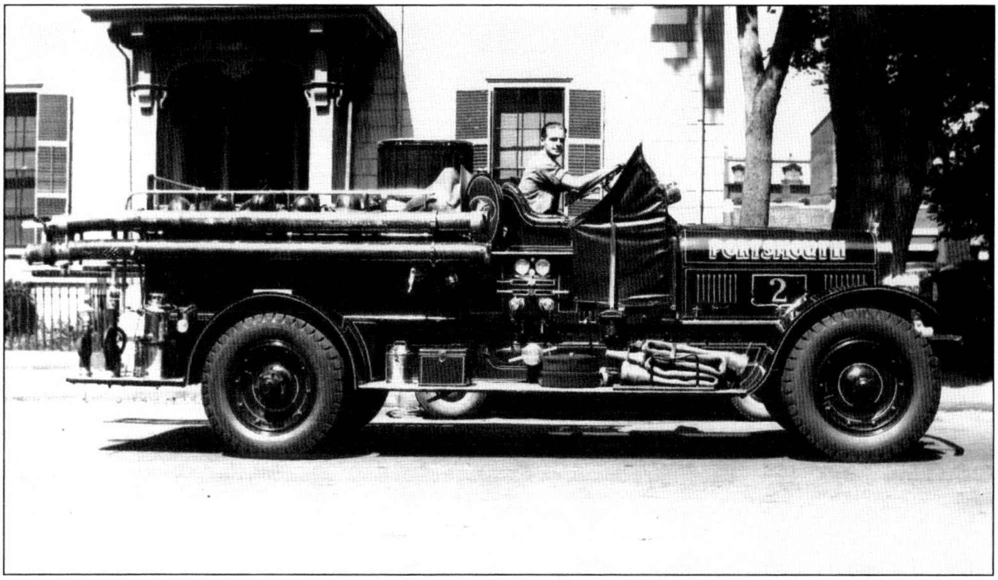

Two more fire engine photographs taken on August 31, 1944, are seen here: Portsmouth engine No. 2 with firefighter Eliot H. Staples (seen above) and engine No. 6 with Ernest W. Weeks (seen below). Firefighter Staples joined the department as a substitute in October 1943 and then as a call fireman in November 1943. He was appointed as a permanent firefighter in March 1945 and would serve as both the first and second assistant engineer of the department. Firefighter Weeks became a call man for Ladder No. 1 in 1931 and a permanent firefighter in 1935. He would also serve as first and second assistant engineer, and in 1967, he was appointed the department's fire chief. (Photographs by Francis J. Carbone; courtesy of Robert Randall.)

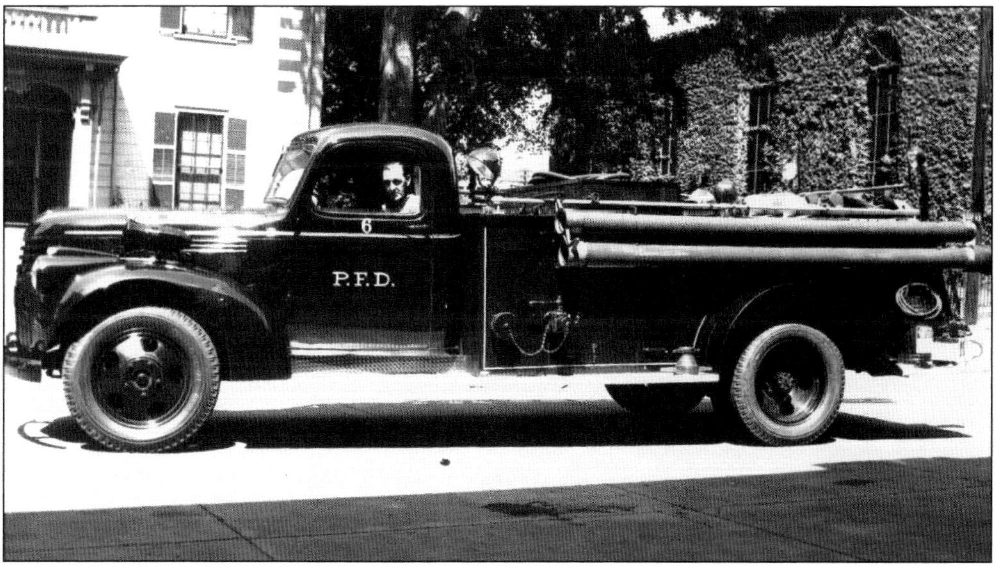

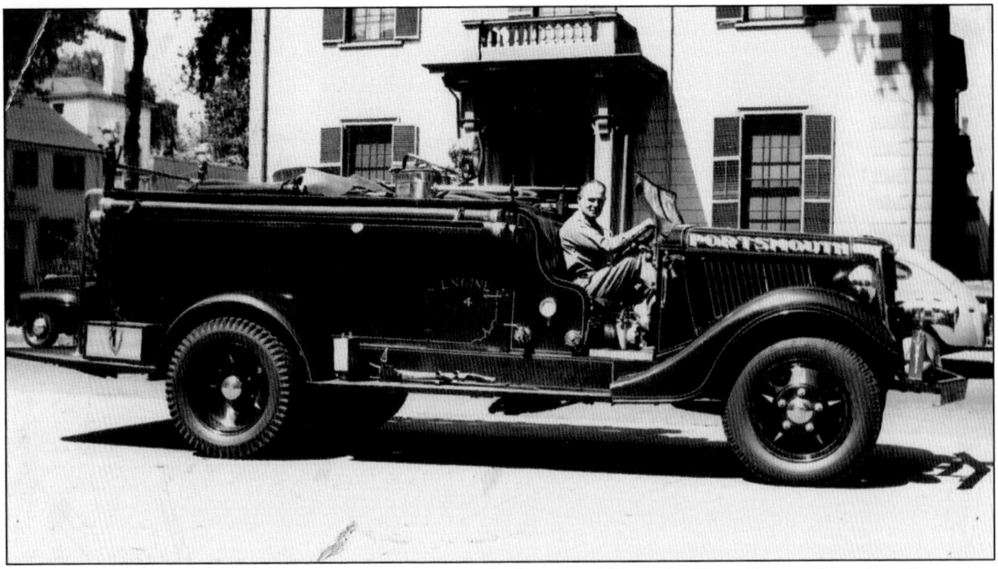

These images of Portsmouth's engine No. 4 with driver William Palfrey (seen above) and engine No. 1 with firefighter Joseph H. Hudson (seen below) were also taken on August 31, 1944. Firefighter Palfrey was a call man assigned to Ladder No. 1 in 1927 and served as a call lieutenant and captain throughout his career. Firefighter Hudson was appointed as a permanent member of the department in 1942 and was promoted to captain in 1954. In 1962, he obtained the rank of second deputy, and five years later, he would rise to first deputy. Deputy Hudson retired from the department in 1973. (Photographs by Francis J. Carbone; courtesy of Robert Randall.)

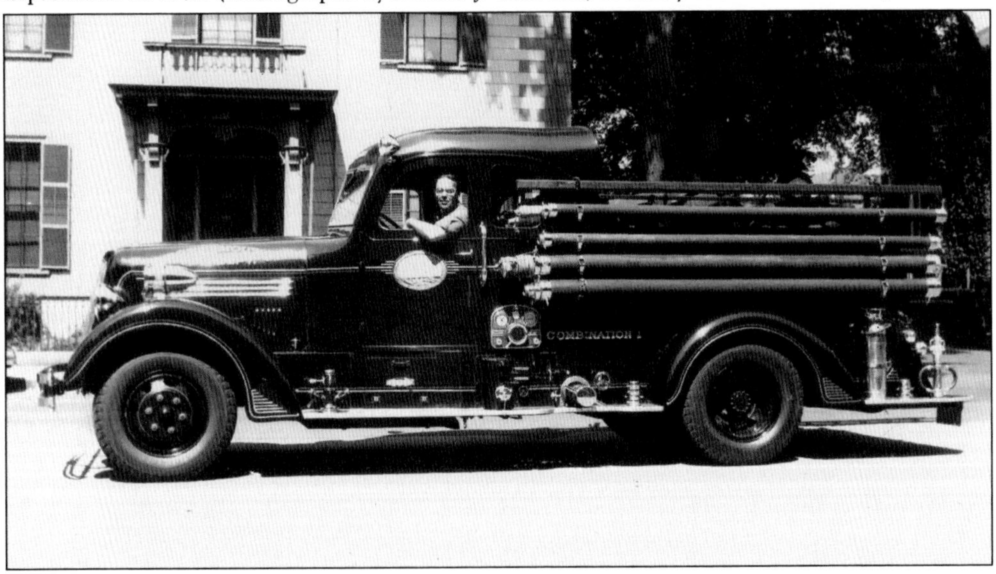

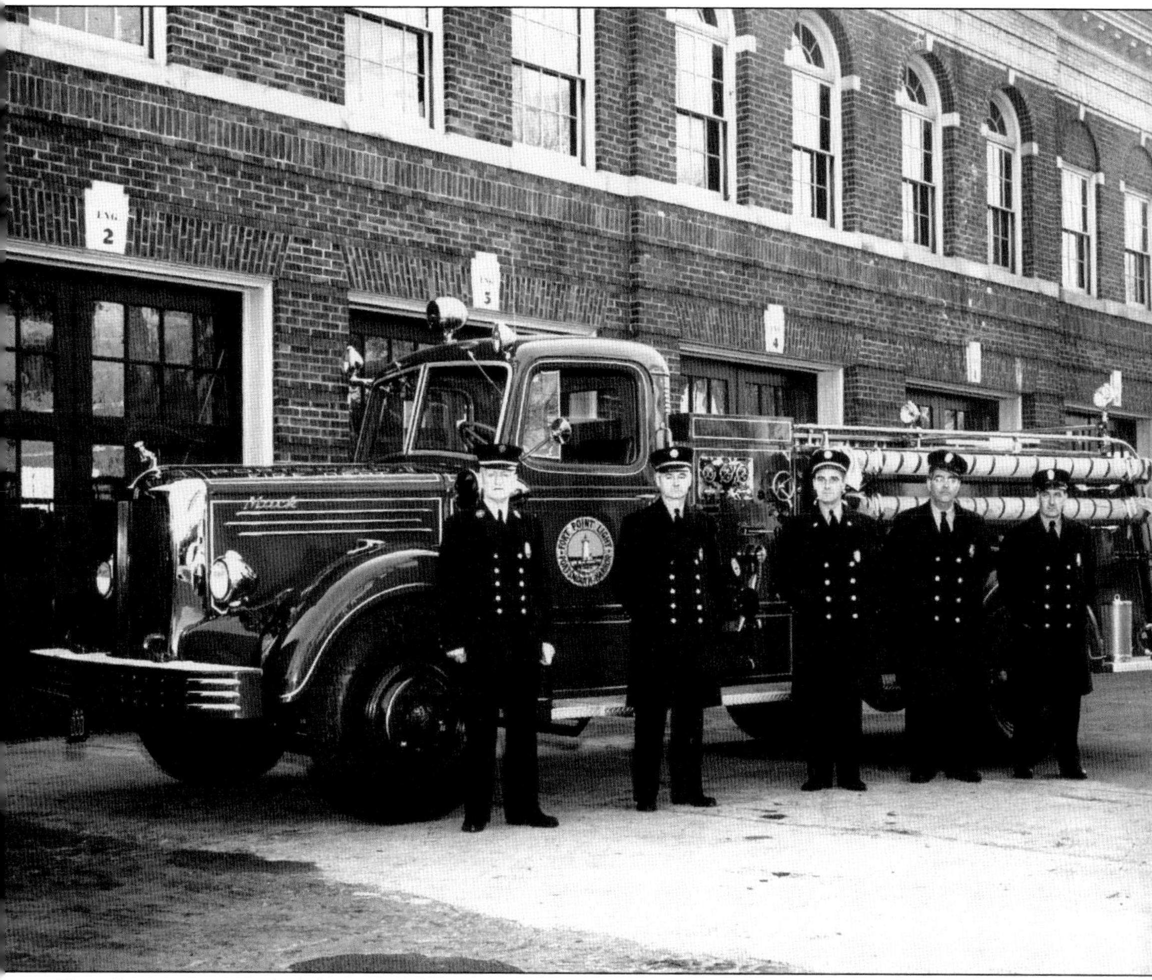

The department took delivery of this Mack triple combination engine on December 19, 1950. The engine, which boasted a 750-gallon-per-minute pump, came fully equipped and included a two-way radio. The purchase was recognized as necessary due to the fact that the latest other pumper had been purchased back in 1932. Looking closely at the image, one can see that the hard suction hoses are still wrapped with protective coverings from the factory, indicating that this photograph was most likely taken on the day the fire engine was delivered. Seen in this photograph taken in front of the Central Fire Station on Court Street are, from left to right, Chief George T. Cogan, First Deputy Chief Fredrick R. "Ted" Crompton, Second Deputy Chief Eliot Staples, driver Joseph Hudson, and firefighter Ernest Weeks.

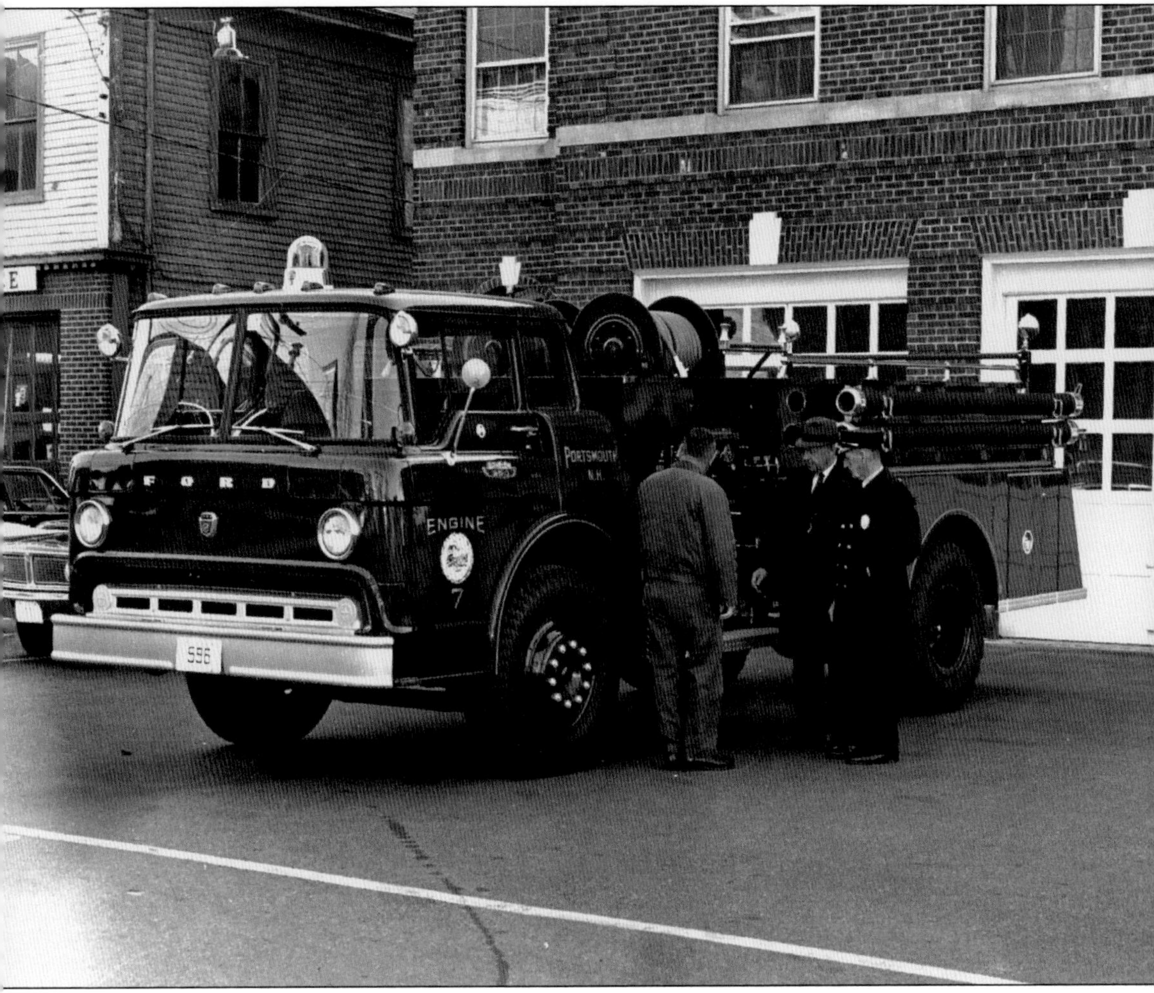

In 1968, the department received a new Ford 750-gallon-per-minute pumper. The new fire engine increased the number of department engines to six; three 750-gallon-per-minute and three 500-gallon-per-minute pumpers. In addition, the department's fleet included one 85-foot ladder truck, one ambulance, one pickup, and one chief's car. The ambulance was a new Superior Coach ambulance on a Pontiac chassis and was delivered to the department in July 1968 at a cost of $10,497. Seen here inspecting engine No. 7 in front of the Central Fire Station on Court Street are, from left to right, two unidentified factory representatives and Fire Chief Ernest Weeks.

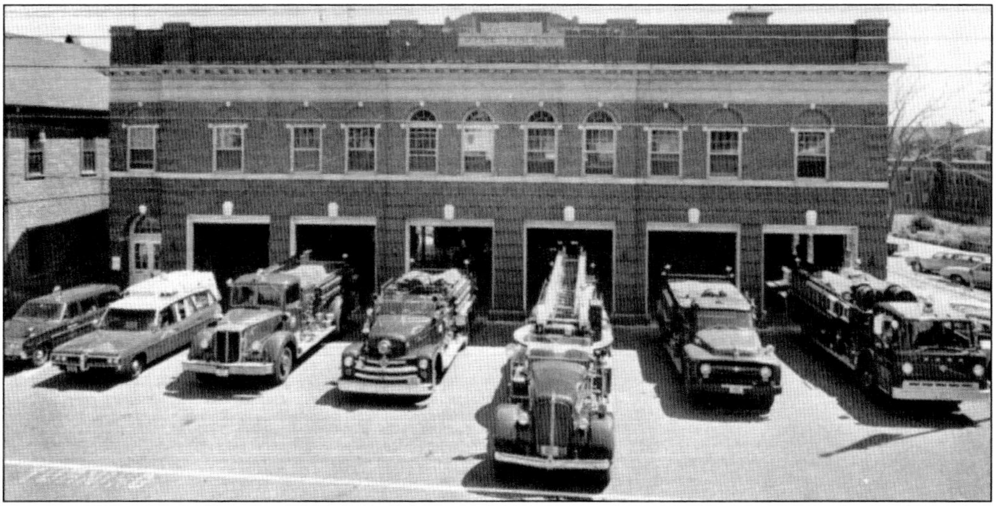

In 1968, the Portsmouth Fire Department boasted a firefighting force of 33 permanent men with 5 additional members added that year, along with 20 call men. The hours of duty for the permanent men were 24 hours on and 24 hours off. This tour occurred over four days and was then followed by three days off. That same year the department responded to 1,347 total calls for service. In this photograph of the Central Fire Station, taken for the 1968 city annual report, are the chief's car, the department's ambulance, four pumpers, and the ladder truck. (Courtesy of City of Portsmouth.)

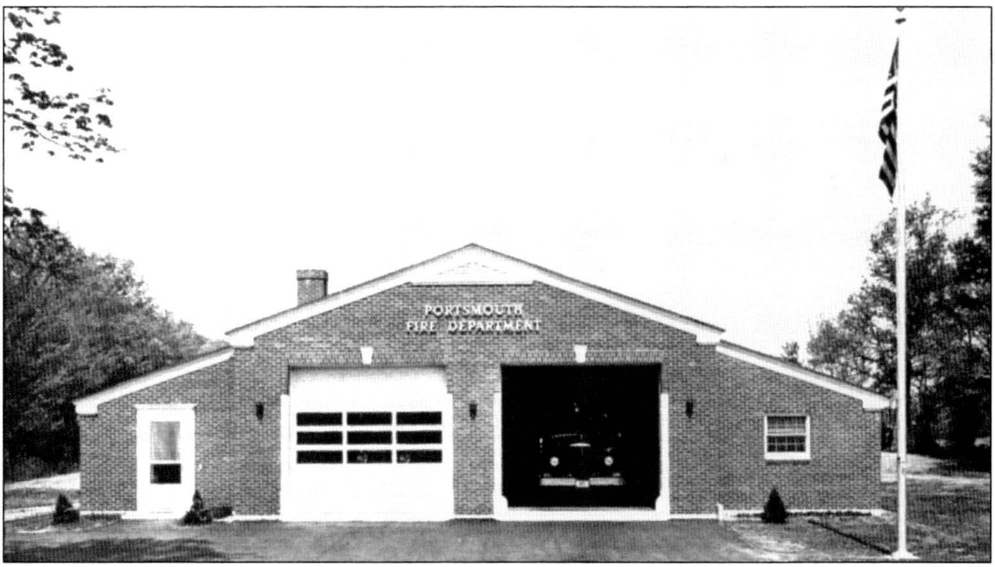

The department's new sub–fire station on Lafayette Road went into service on September 25, 1967. Both engines No. 3 and 5 were transferred from the Central Fire Station to this new firehouse. The station was manned by two shifts of one lieutenant and four privates each. The station was dedicated on October 7, 1967. The flagpole to the right of the station was donated by Alice Marden in memory of former Portsmouth fire chief Herbert Marden (1903–1904). (Courtesy of City of Portsmouth.)

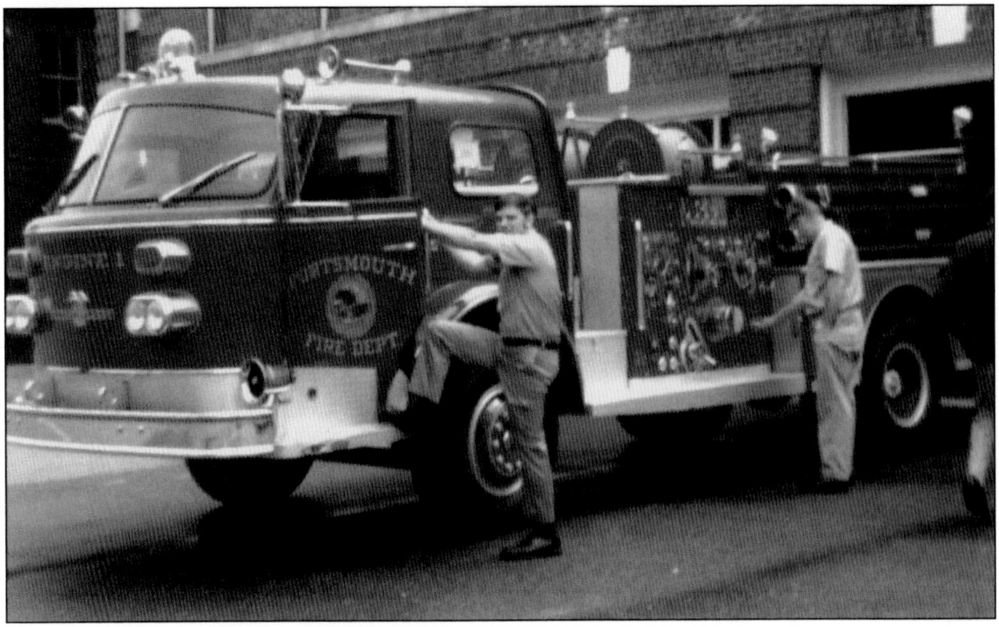

In 1973, an American LaFrance 1,500-gallon-per-minute pumper was delivered to the department. After delivery, it was found to have been in a flood and therefore returned to the company. In this photograph, firefighters Stanley Whitney (left) and Baylor Horvath (right) are performing their morning checks on the new American La France 1,000-gallon-per-minute pumper that arrived in the spring of 1974.

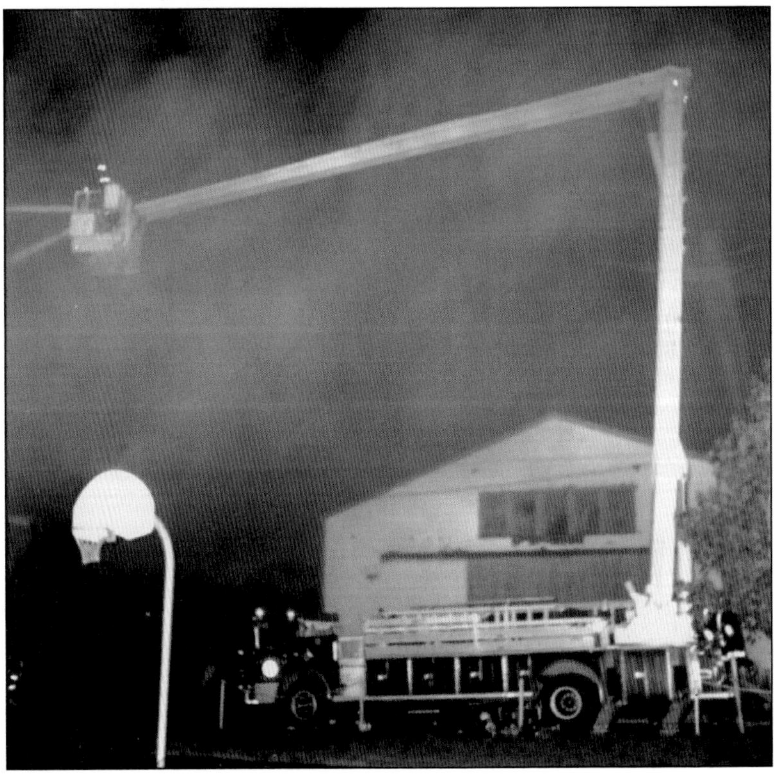

On August 1, 1970, the department received a new 80-foot open-cab American LaFrance Aero Chief articulating boom snorkel truck. It is seen here operating at a Rock Street fire on July 25, 1977. The department's Aero Chief was one of 171 built between 1962 and 1974 by the American LaFrance Company. Portsmouth's Aero Chief was in service until 1994.

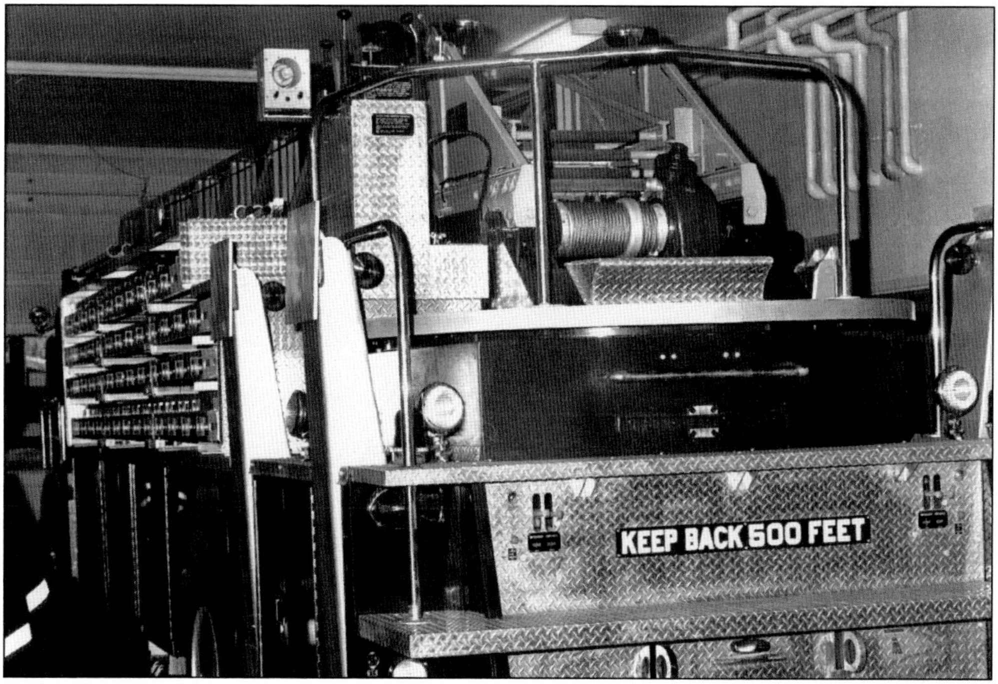

Ordered in October 1975 to have a four-section, steel aerial ladder that would be hydraulically operated, this 100-foot American LaFrance aerial ladder was delivered to the department at a cost of $133,500 in December 1977. This photograph of the ladder truck was taken inside Station No. 2 on Lafayette Road on September 23, 1981.

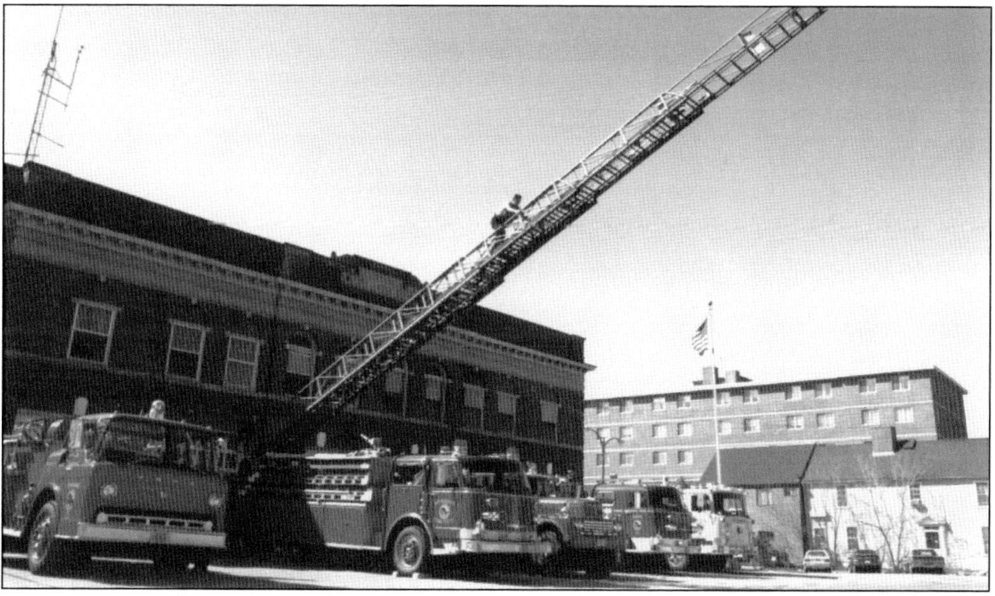

Capt. Everett "Al" Weare climbs the aerial of ladder No. 1 in this March 1986 image of the Central Fire Station on Court Street. The fire apparatus in the photograph are, from left to right, the 1968 Ford pumper, the 1977 American LaFrance aerial ladder truck, the 1965 Howe pumper, the 1974 American LaFrance pumper, an the 1982 Mack pumper. A year earlier the department would respond to 1,441 fire calls, including 13 working fires, and 1,221 ambulance calls.

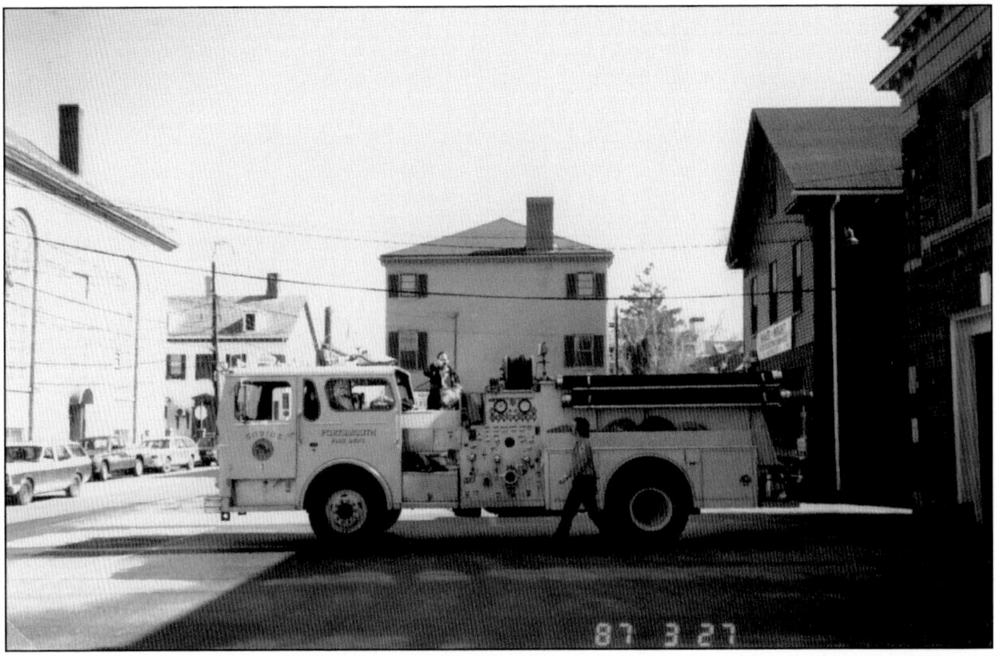

In this photograph taken at the Central Fire Station on March 27, 1987, firefighters Ricky A. Plummer and Richard B. Duddy are performing their morning checks on the department's 1980 Maxim pumper. During that same month in 1987, the creation of the Portsmouth Fire Commission was approved by city vote, and in 1988, the first commissioners were appointed by the mayor. After 1990, the citizens of the city would elect the members of the Portsmouth Fire Commission. The commission, still in existence today, is charged to work on policy and administrative matters within the department, being a level between fire chief and city council.

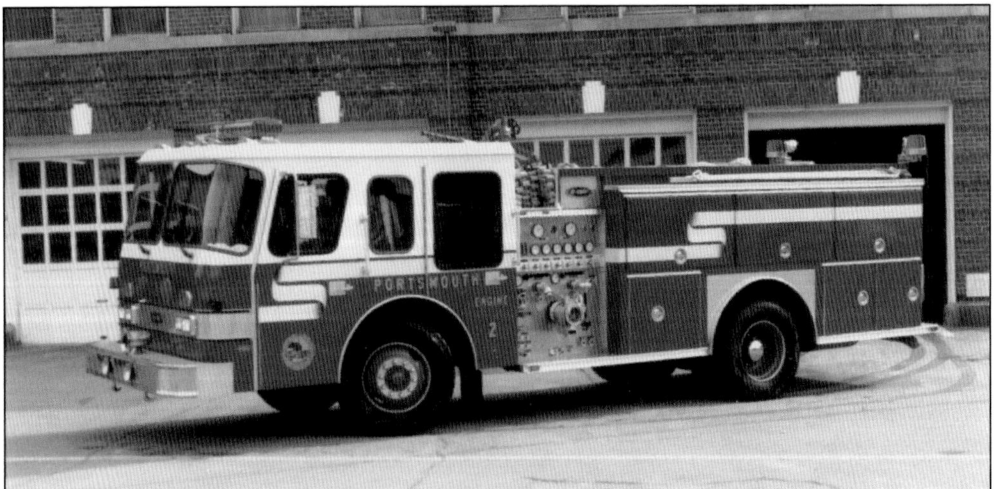

In 1985, Emergency One introduced the Hush rear-engine chassis fire pumper. In 1987, the Portsmouth Fire Department took delivery of a 1,500-gallon-per-minute Hush pumper. The engine was one of five pumpers serving the department during the late 1980s. Other department apparatus included two aerials, two ambulances, one fire alarm truck, and three utility vehicles. This photograph of engine No. 2 was taken in front of the Central Fire Station on February 19, 1988.

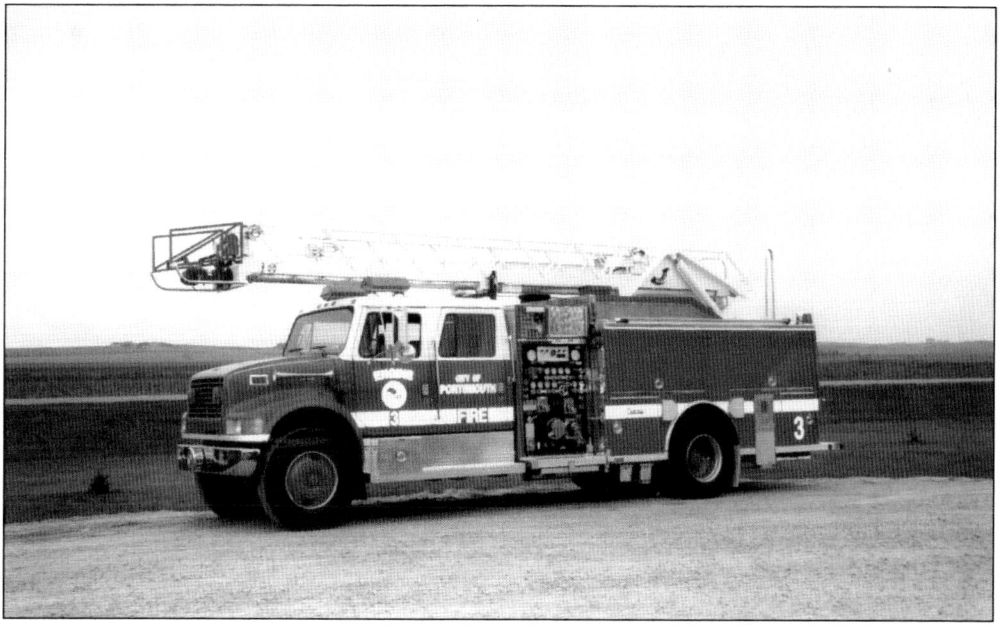

The department opened Station No. 3 at the former Pease Air Force Base's structural fire fighting station in 1993. In addition to adding more firefighters to the force in order to properly staff the new station, the department purchased two Central States pumper-aerials on International 4900 chassis. Known as "quints," the fire engines were characterized by five features, a fire hose, a pump, a water tank, ground ladders, and an aerial device.

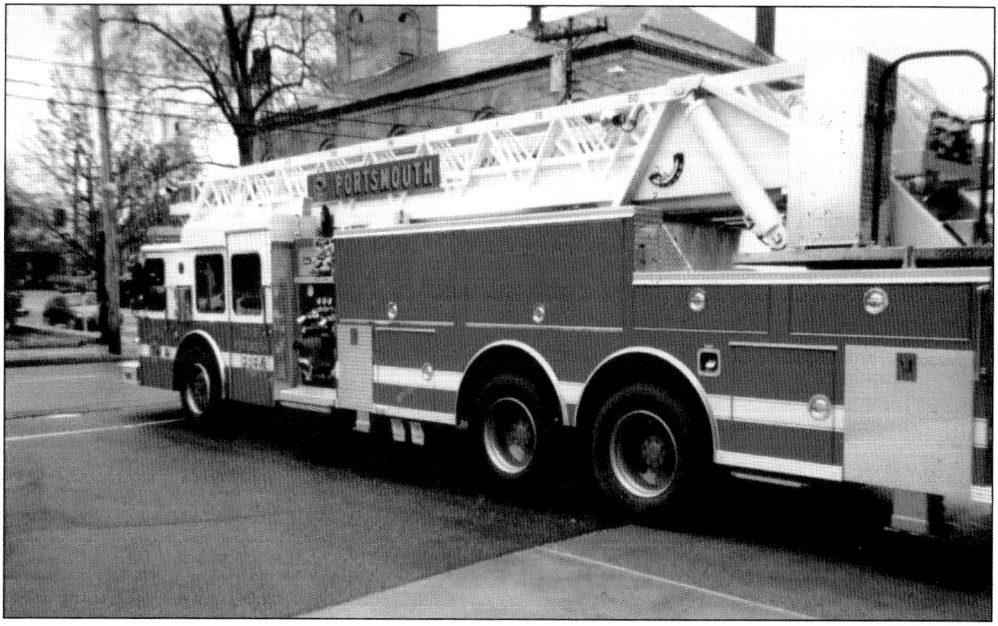

In 1996, the Portsmouth Fire Department replaced its 100-foot American LaFrance ladder truck with a new 110-foot aerial that also boasted a 1,500-gallon-per-minute Hale fire pump. The new ladder No. 1 was built by Central States Fire Apparatus on an HME chassis. Throughout its 11 years of service, the truck would also be known as ladder No. 2 and ladder No. 5, running out of both Station No. 1 and Station No. 2.

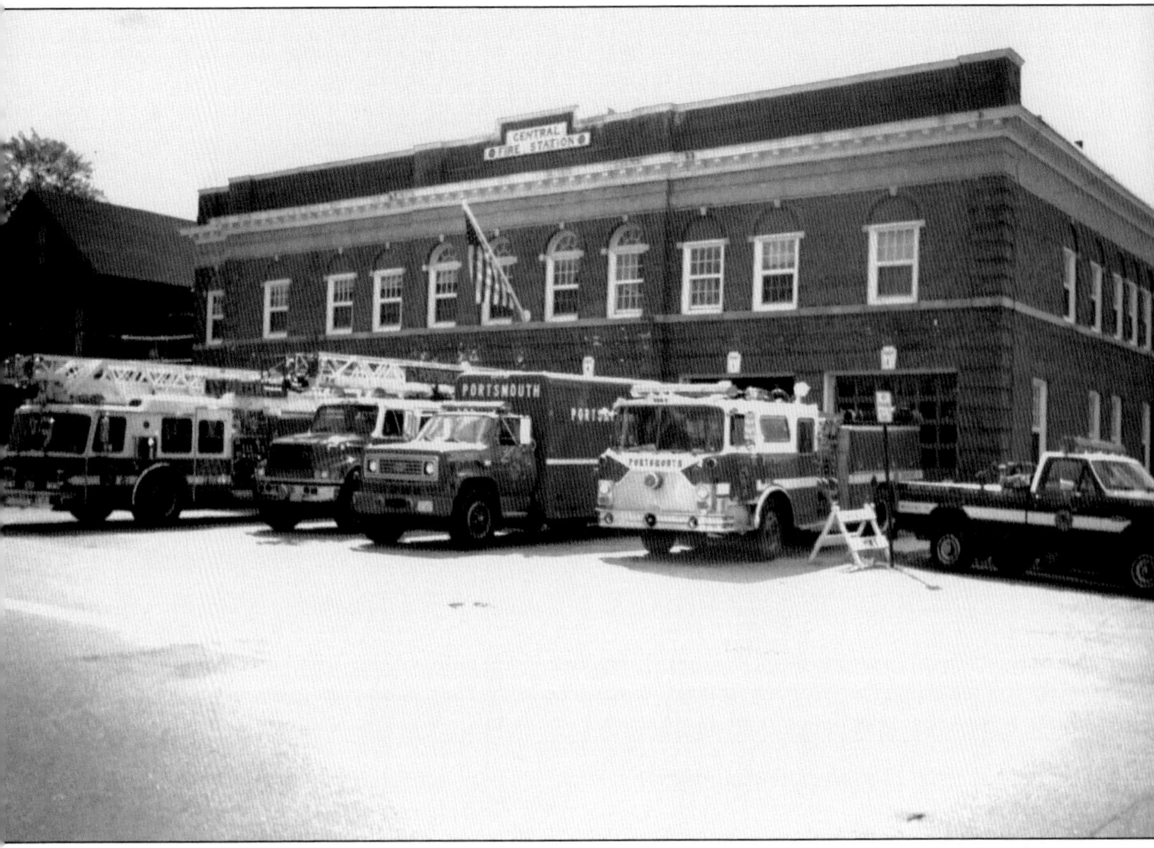

Sometime after Station No. 3 opened, the Central Fire Station at 170 Court Street became known as Station No. 1. The city's fire apparatus is on display in this image from 1997. During this same year, renovations to the second floor of the station eliminated the large hall and created individual bedrooms, while also updating utilities, environmental systems, and living spaces.

Three

FIREFIGHTERS

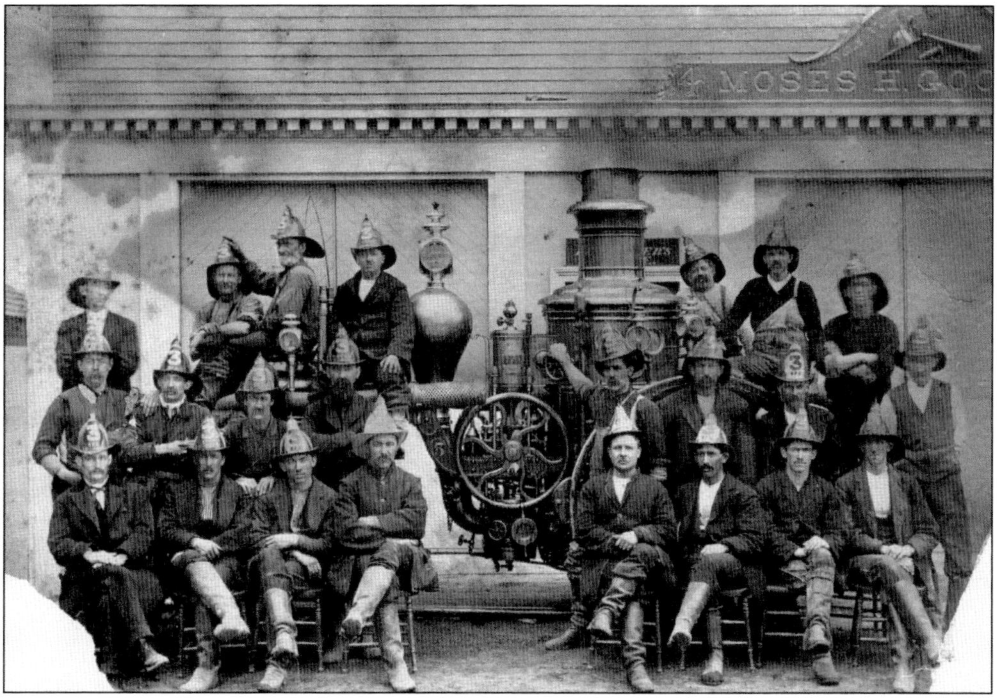

The fire companies and engine houses of the department in 1876 were Dearborn SFE No. 1 and Moses H. Goodrich SFE No. 4 on Hanover Street; Sagamore SFE No. 2 on lower Court Street; Kearsarge SFE No. 3 and the Garibaldi Hook and Ladder Company at the Central Engine House on Court Street; Extinguisher Company No. 5 on State Street; Extinguisher Company No. 6 on Elm Street; two hand engines at the arsenal on South Street (near South School Street and Johnson's Court); and one hand engine at the city farm on Myrtle Street. In this photograph taken around 1876, in front of the Central Engine House, is the Kearsarge SFE Company No. 3. (Courtesy of the Connors family.)

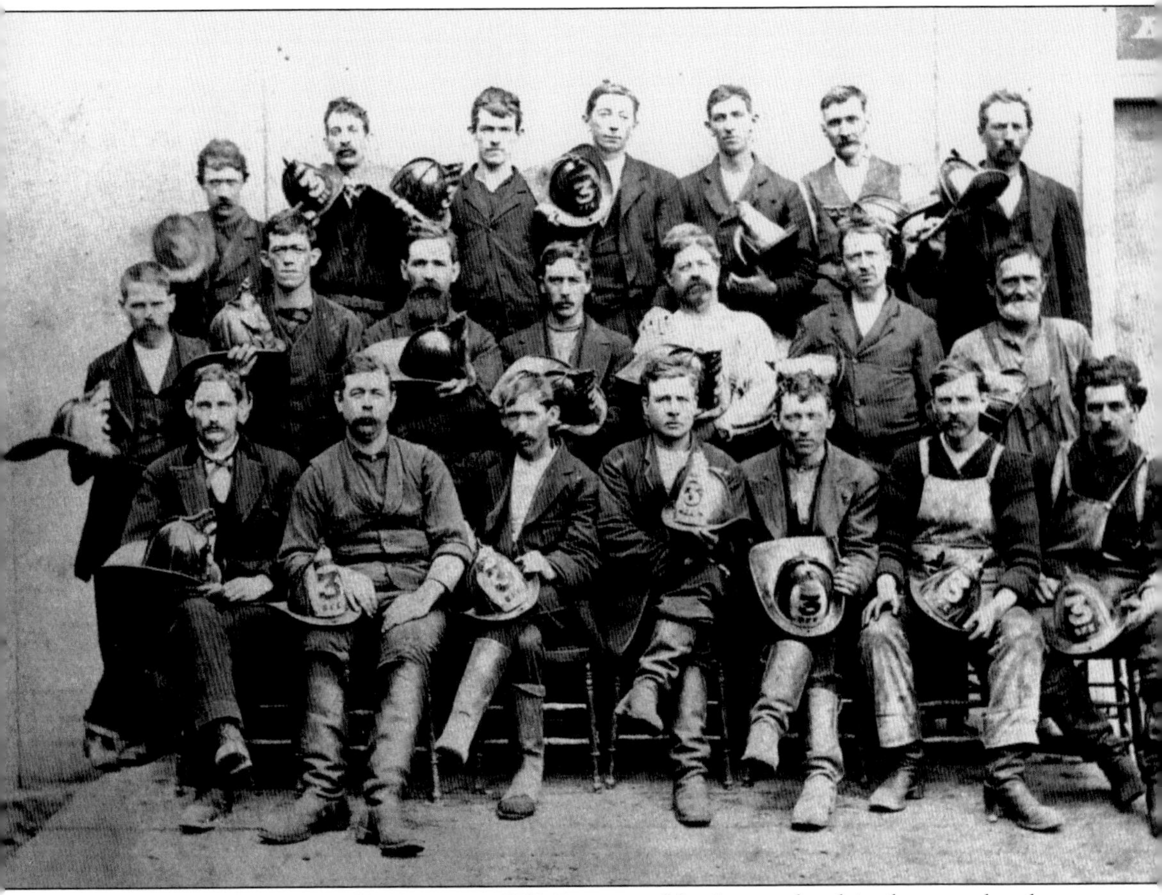

Members of Portsmouth's Kearsarge SFE Company No. 3 pose for this photograph taken on May 8, 1879. Seen here are, from left to right, (first row) Robert Phinney, James Hurd, Harvey Lear, John Harrison, Charles Twombly, Charles Brock, and George N. Jones; (second row) George Smart, ? O'Donnell, Frank Seavey, Charles Coleman, George Raitt, Ben Atwell, and Enoch Jones; (third row) Walter Cate, John Clark, ? O'Donnell, Gil Hayes, Frank Hayes, Herbert Sheldon, and John Moulton. (Courtesy of Strawbery Banke.)

J. Frank Shannon served as chief engineer from 1890 to 1891. He began as a call firefighter with the Moses H. Goodrich SFE Company No. 4 on September 1, 1890. Later he would be a member of Colonel Sise Company No. 2. Shannon was honorably discharged from the department on December 1, 1903. (Courtesy of Robert Randall.)

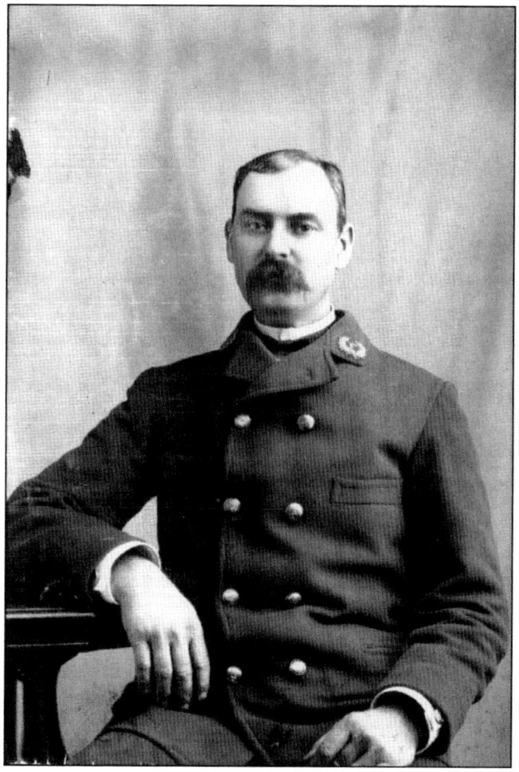

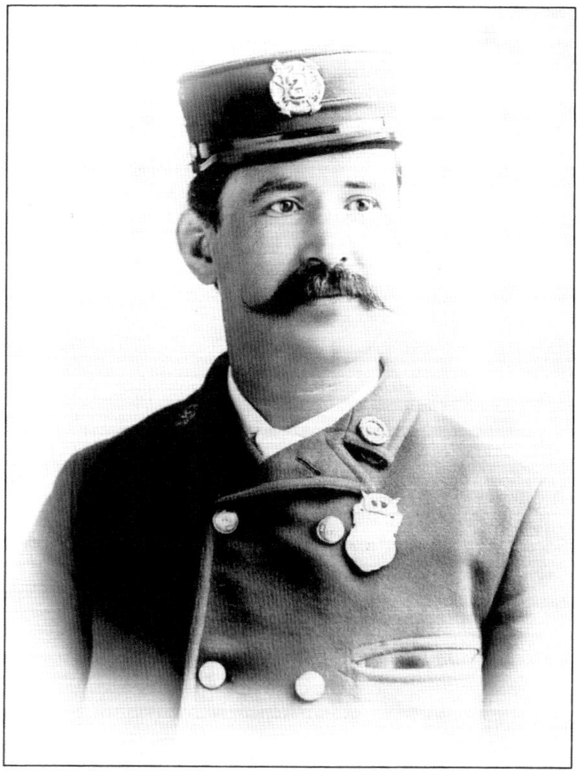

By 1899, the roster of the department would consist of 1 chief engineer, 4 assistants, 1 superintendent of fire alarm, 6 permanent men, and 93 call firefighters. Eugene Sullivan, seen in this portrait, was a call firefighter beginning in April 1899. He would serve as chief engineer during 1899 to 1900 and would leave the department in 1901.

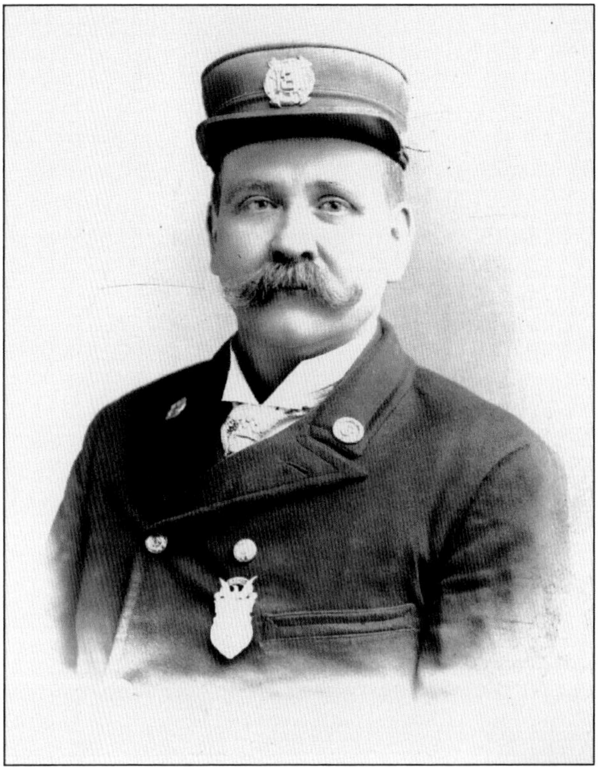

In 1886, the Portsmouth Fire Department boasted a complement of 123 firefighters operating four steamers, three hose wagons, one carriage, one hook and ladder, and two sleighs. This is a portrait of Libby Bennett, who served as a call firefighter for the Kearsarge SFE Company from May 3, 1886 to February 1, 1887.

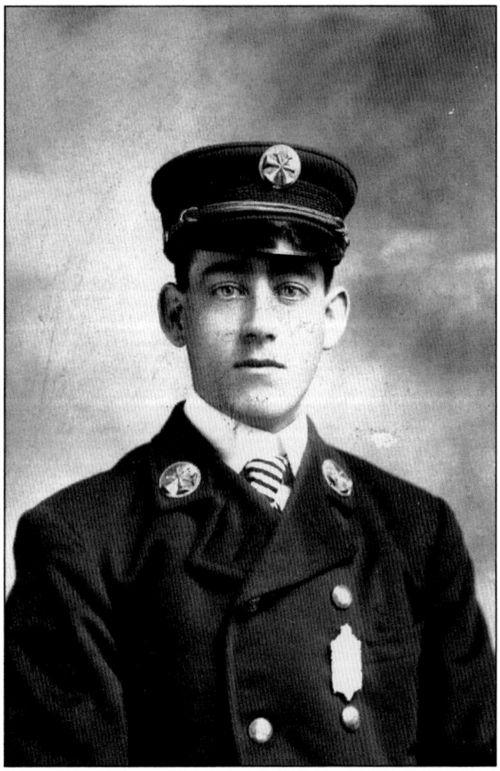

At some point between 1916 and 1924, this portrait was taken of Perley D. Hersey. His collar and hat insignias represent the rank of assistant engineer, which may indicate that this photograph was taken closer to 1924. Hersey was a member of the Colonel Sise Company beginning in 1908 and, after serving on the board of engineers, returned to the company, reaching the rank of captain. He resigned on January 6, 1943. The records of the department also indicate that Hersey served in the U.S. Army in 1917 and 1918.

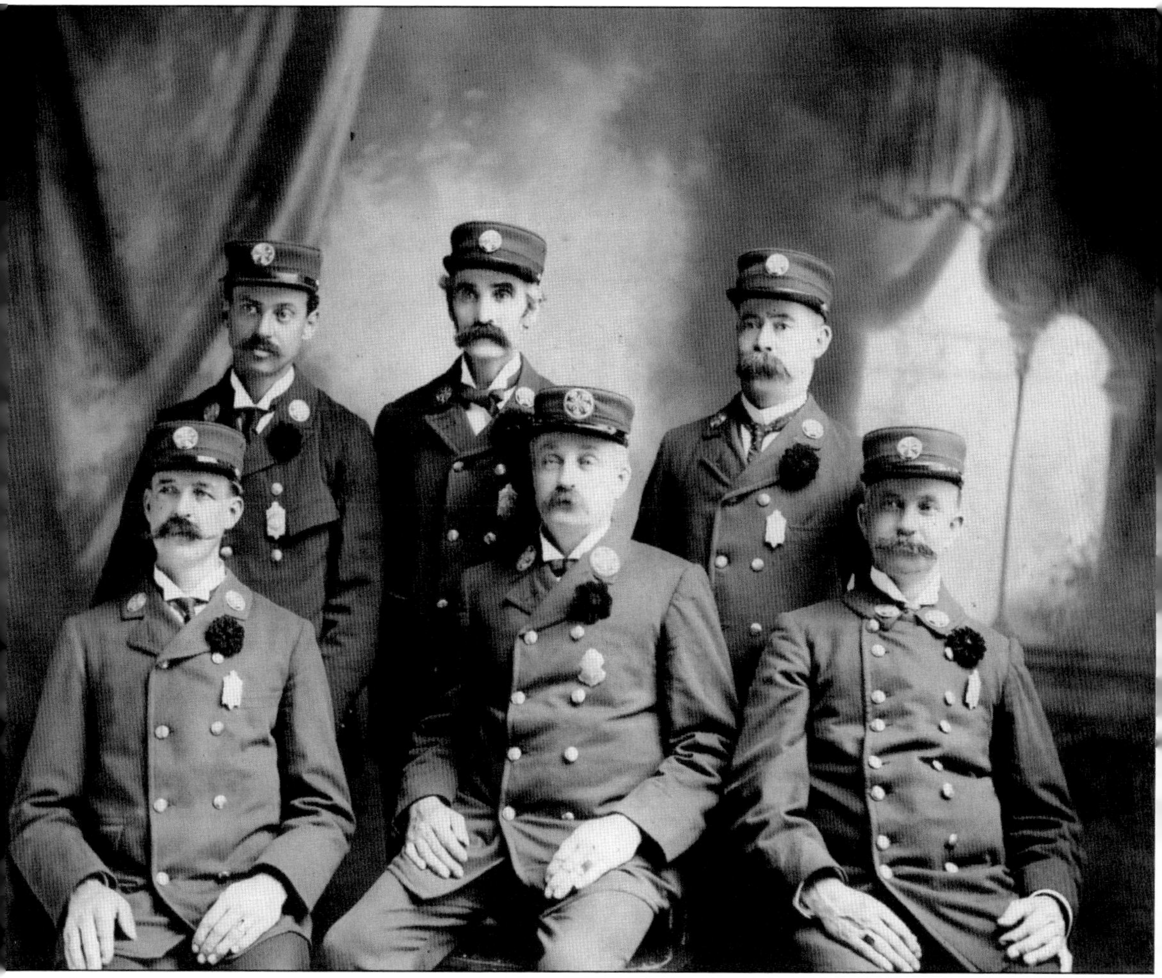

The five-member board of engineers and clerk pose for this photograph taken in 1903. Seated in the middle of the first row is Chief Engineer Herbert Marden. On November 2, 1874, Chief Marden joined the Kearsarge SFE Company No. 3. He served as chief engineer both from 1884 to 1885 and 1903 to 1904. He ended his tenure with the department as a call man for Kearsarge SFE Company No. 3 on July 1, 1912. The clerk of the board, Robert M. Herrick, is standing in the back row on the left. He was a member of the Willard J. Sampson Hook and Ladder Company.

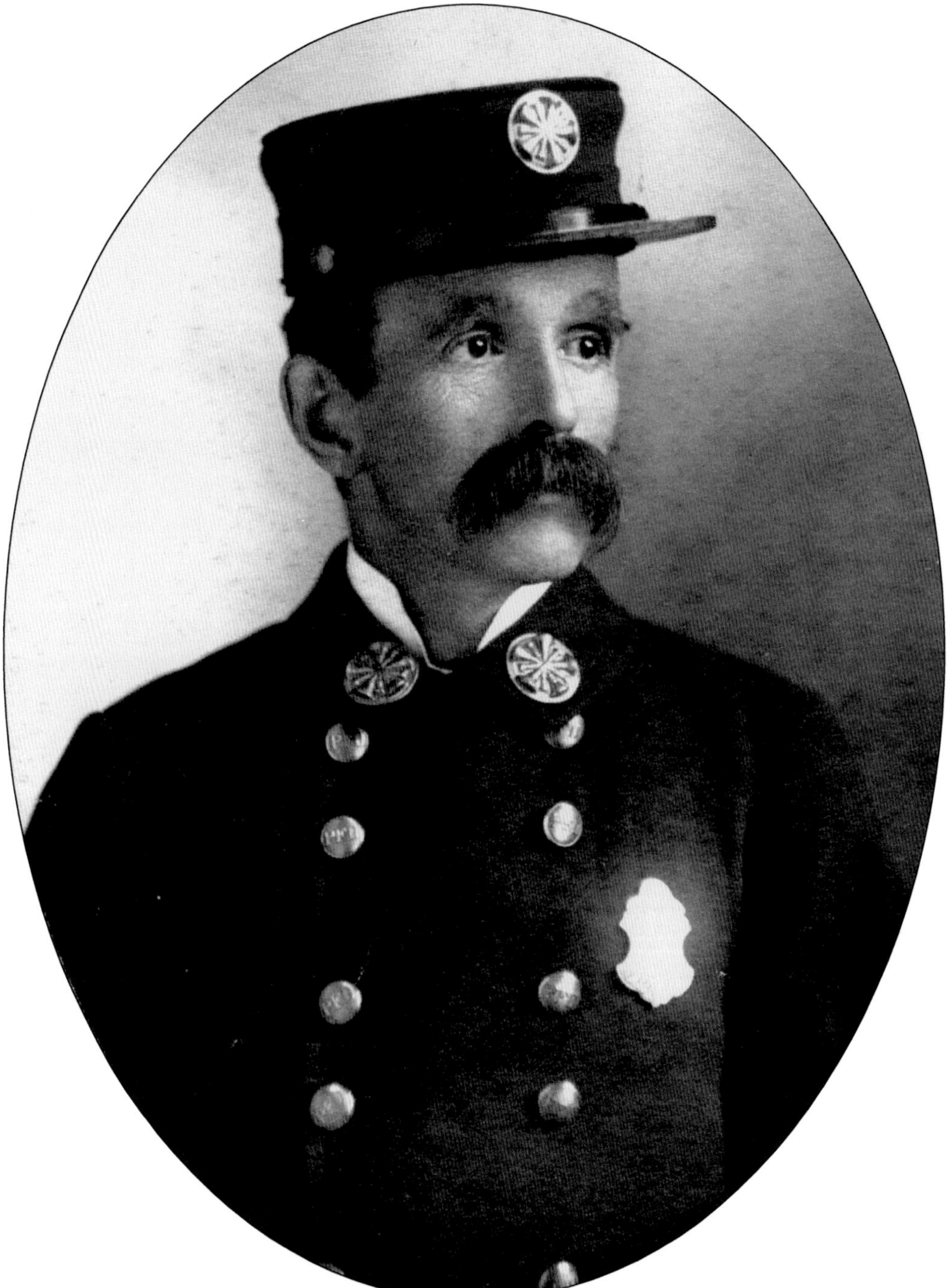

John Dame Randall began his career with the department on January 6, 1886, as a call firefighter with Kearsarge SFE Company No. 3. He would serve as the chief engineer from 1895 to 1903 and again in 1905. He would hold the position again for three more years prior to his honorable discharge from the department in 1912. Chief Randall was instrumental in modernizing the department when he championed the purchase of Portsmouth's first gasoline-powered fire engine in 1912. (Courtesy of Robert Randall.)

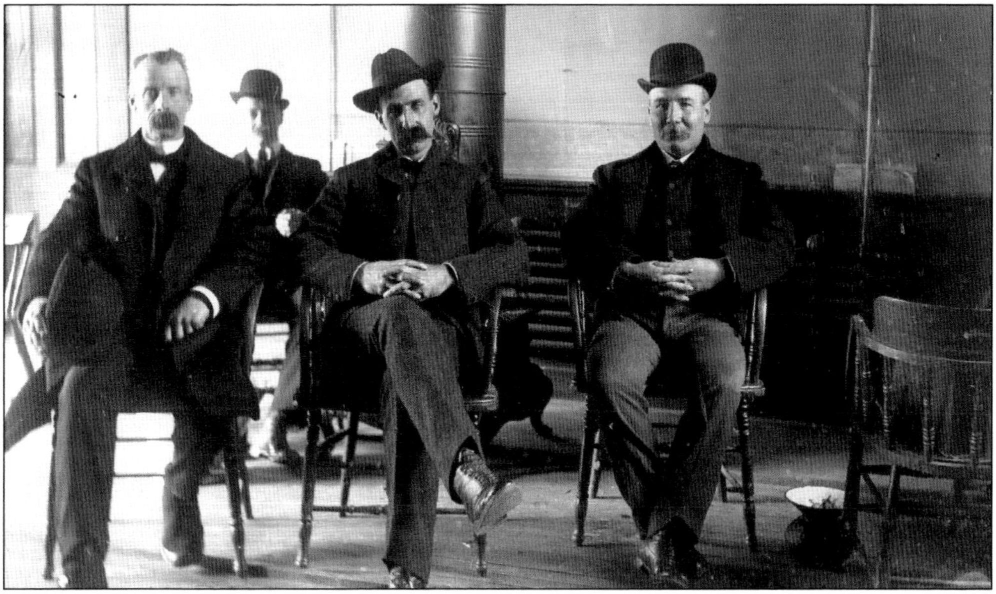

In this undated photograph taken inside the Hanover Street station are, from left to right, Harry ?, unidentified, Josiah Morrison Varrell, and Gus ?. Varrell was a call firefighter starting on January 1, 1890. He would reach the rank of both lieutenant and captain with the Moses H. Goodrich SFE Company No. 4, as well as serve on the board of engineers. During 1916, he was assigned to Colonel Sise SFE No. 2 but would soon be transferred back to Moses H. Goodrich SFE Company No. 4. Varrell was honorably discharged from the department on September 7, 1921.

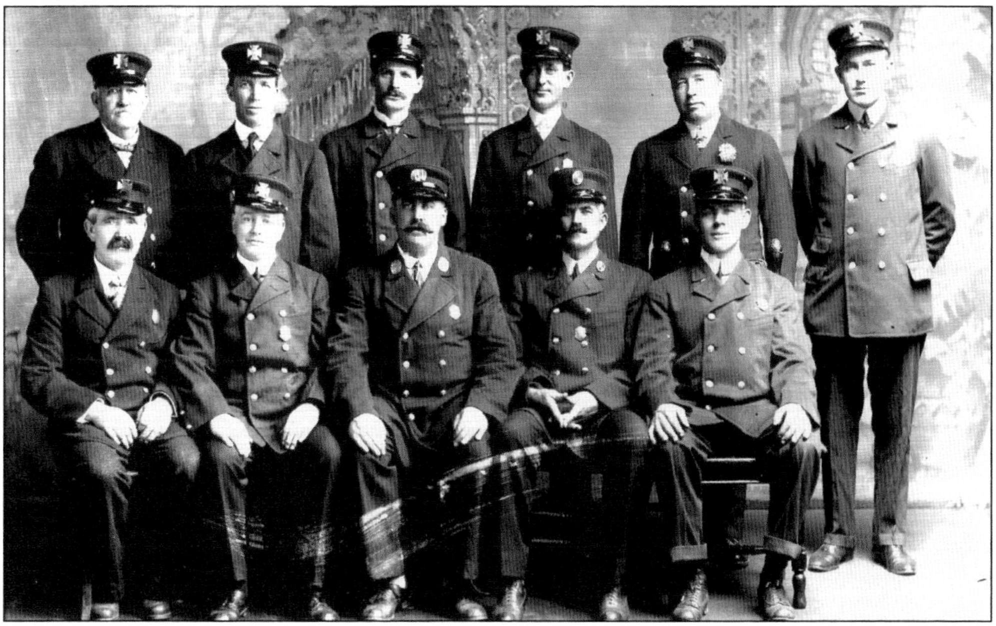

By 1900, a new engine house was built at the corner of Bartlett and Pine Streets. Housed in reserve was the Sagamore SFE Company. The 11 call firefighters of the company in this photograph taken around 1916 are, from left to right, (first row) Tom Fullam, Tom A. Moran, Maurice O'Leary, Daniel Scott, and Frank P. Clair; (second row) William Clapp, James McCabe, Jeremiah Regan, Bill Pendergast, John Quirk, and Dennis Kelly. (Courtesy of Strawbery Banke.)

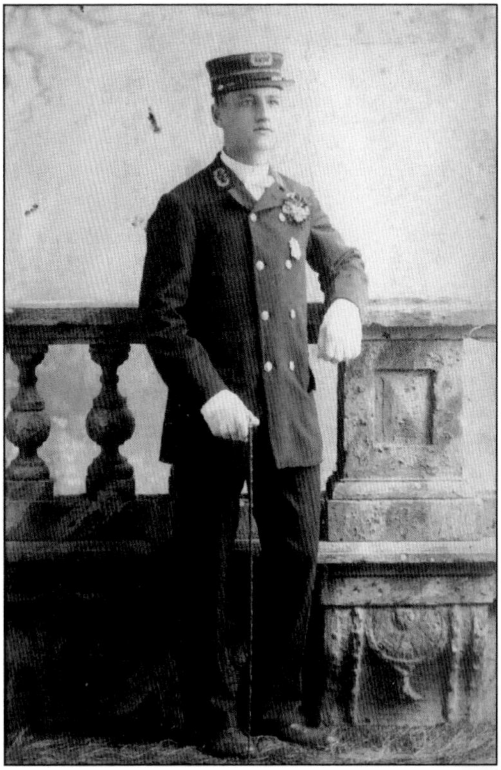

William F. Woods joined the Portsmouth Fire Department as a call man on February 2, 1891. In 1898, he held the rank of lieutenant with the Willard J. Sampson Hook and Ladder Company, and two years later, he was promoted to the position of captain with the same company. From 1906 to 1913, he served on the board of engineers, becoming the call chief engineer from 1913 to 1927. Chief Woods has the distinction of being the first permanent chief engineer of the department and served in that capacity from April 1, 1927, until his death on April 28, 1938.

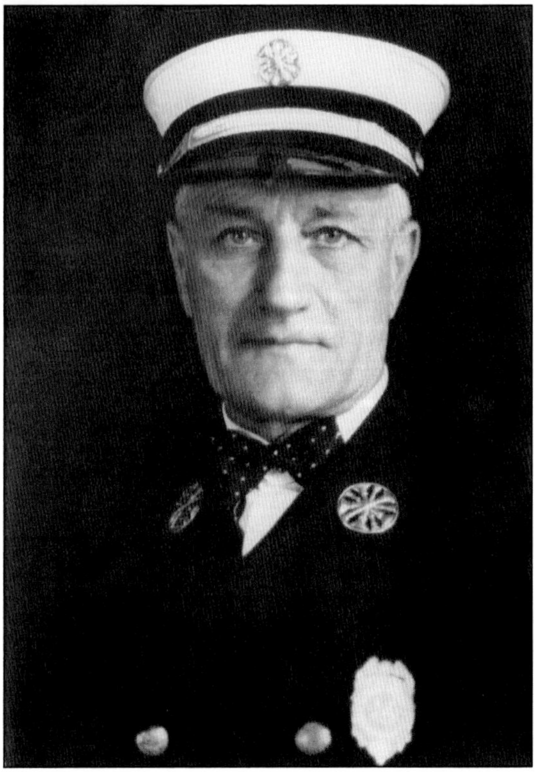

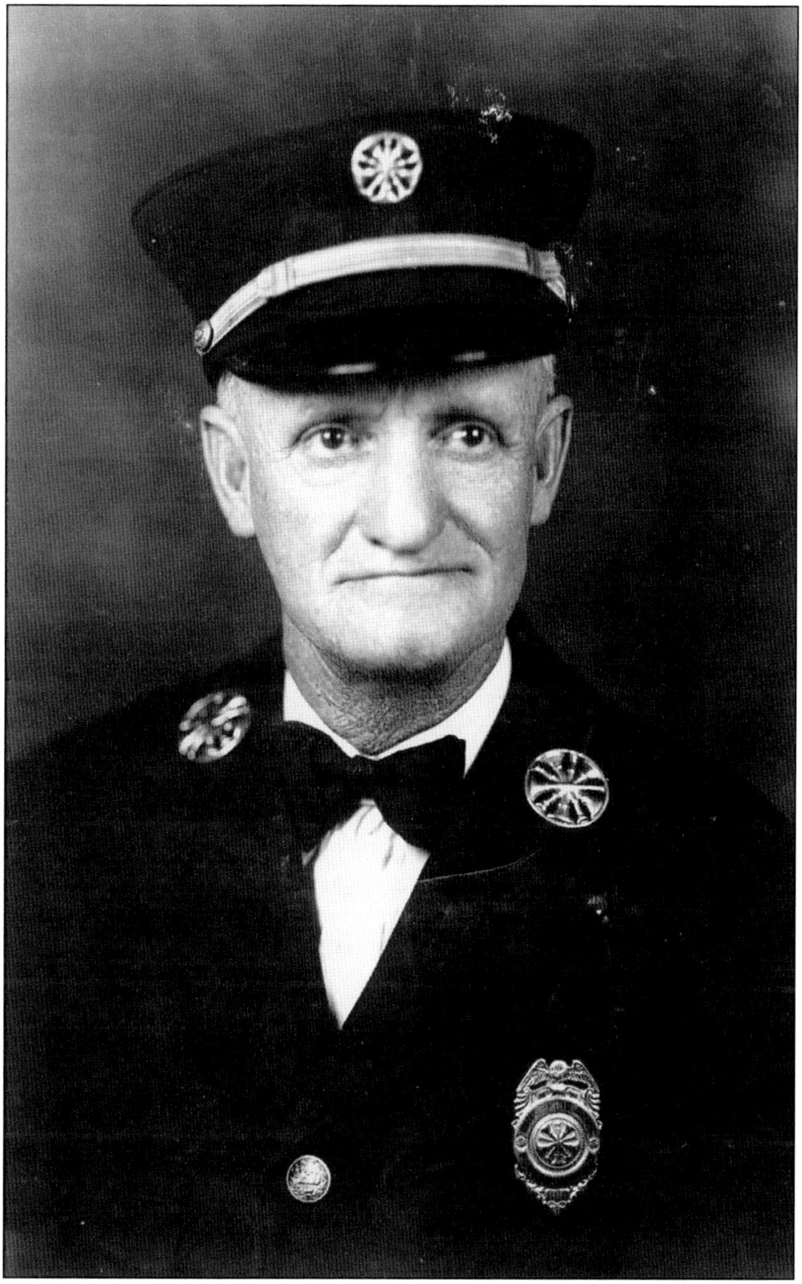

Chief George T. Cogan began his career with the Portsmouth Fire Department as a call man assigned to the Sagamore SFE Company on March 12, 1900. Four years later, he would be appointed as a permanent firefighter. In 1937, he was promoted to first assistant engineer, followed by chief engineer in 1938. The last to have the title of chief engineer, he would retire on October 15, 1952, with a pension of $104.16 a month. In addition to his contributions in modernizing the department, Chief Cogan, in his last will and testament, established a scholarship program to be awarded to "worthy boys of Portsmouth, New Hampshire; graduates of Portsmouth High School or St. Thomas Aquinas of Dover to further their education at recognized universities, colleges or other institutions of higher education."

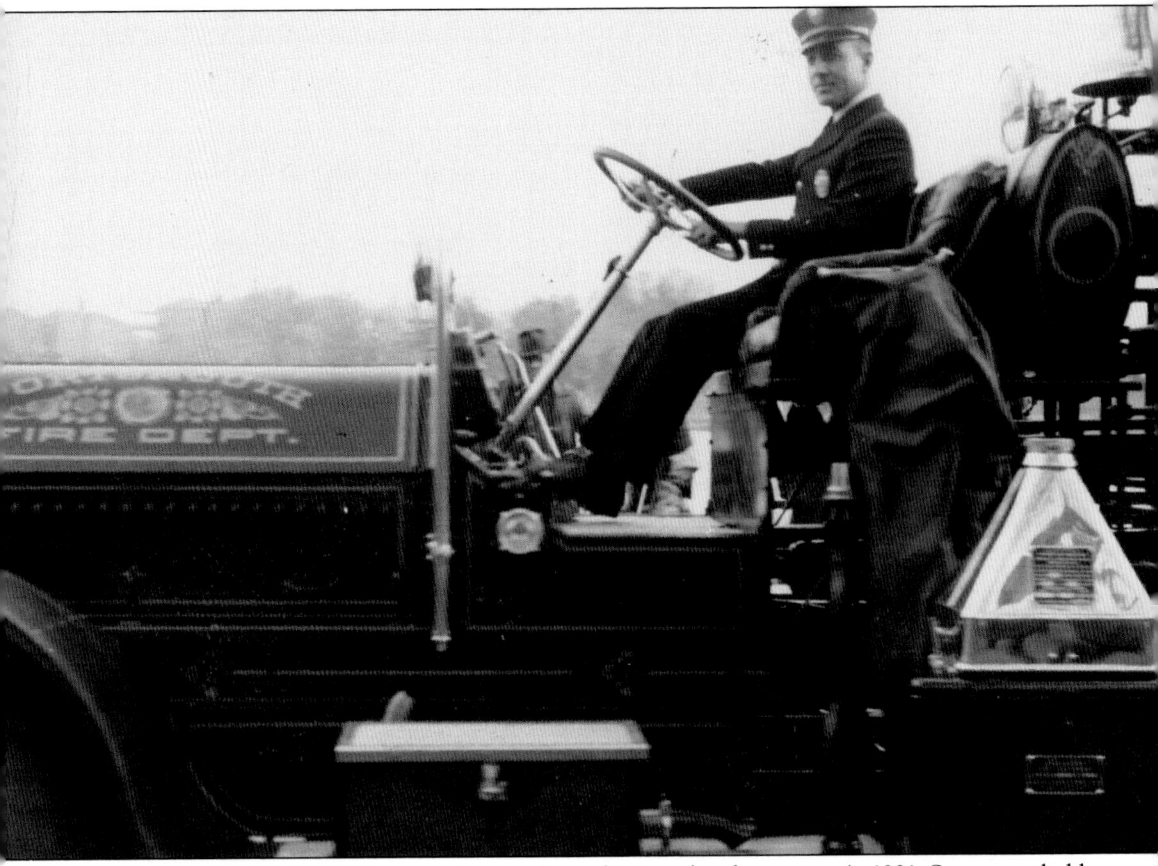

Firefighter John "Bill" Connors is seen here driving the department's 1921 Seagraves ladder truck in this photograph taken in 1940. Firefighter Connors joined the department as a call man in 1937 and was appointed as a permanent firefighter in 1938. Firefighter Connors left the fire department in 1944 to serve in the U.S. Navy during World War II. In 1948, he was again appointed to the position of permanent firefighter with the Portsmouth Fire Department and retired in 1976. (Courtesy of the Connors family.)

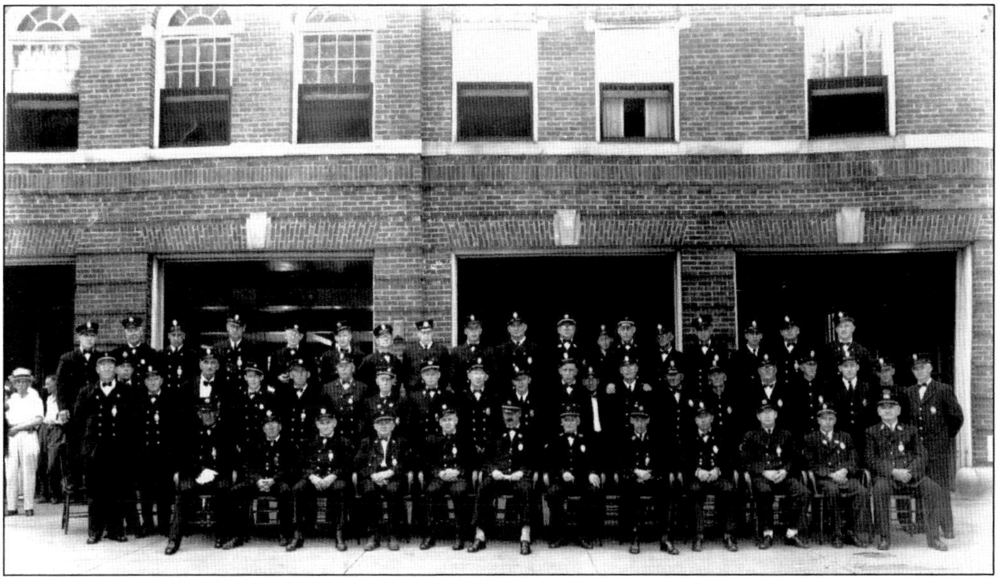

Salaries for members of the fire department in 1935 were $2,100 for the chief engineer, $200 for each of the assistant engineers, $1,700 for each of the eight permanent men, $105 for eight of the call firefighters, $100 for 36 of the call men, and $25 for the clerk. In this image taken sometime around 1939, both call and permanent firefighters, along with the chief officers, pose in front of the apparatus doors of the Central Fire Station. (Courtesy of the Connors family.)

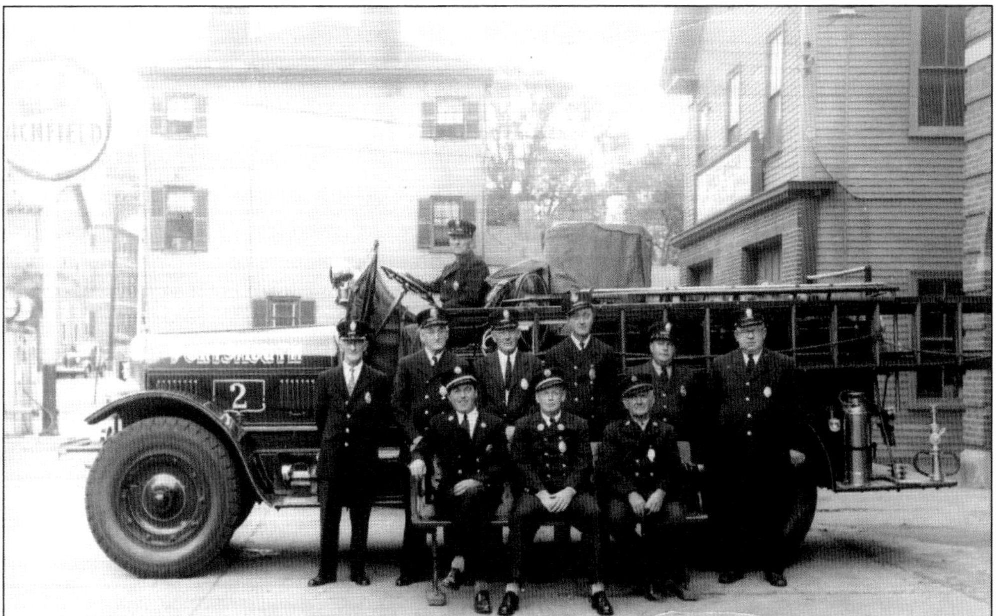

This image of engine No. 2 and the call firefighters assigned to the company was taken in front of the Court Street station on September 7, 1948. In the photograph are, from left to right, (first row) Lt. John Golter, Lt. Robert P. Murphy, and Lt. Frank Dyer; (second row) John "Chick" Morrill, William Drew, unidentified, Forrest "Red" Webb, Alvah Card, and unidentified; (third row) permanent driver Frank Amazeen.

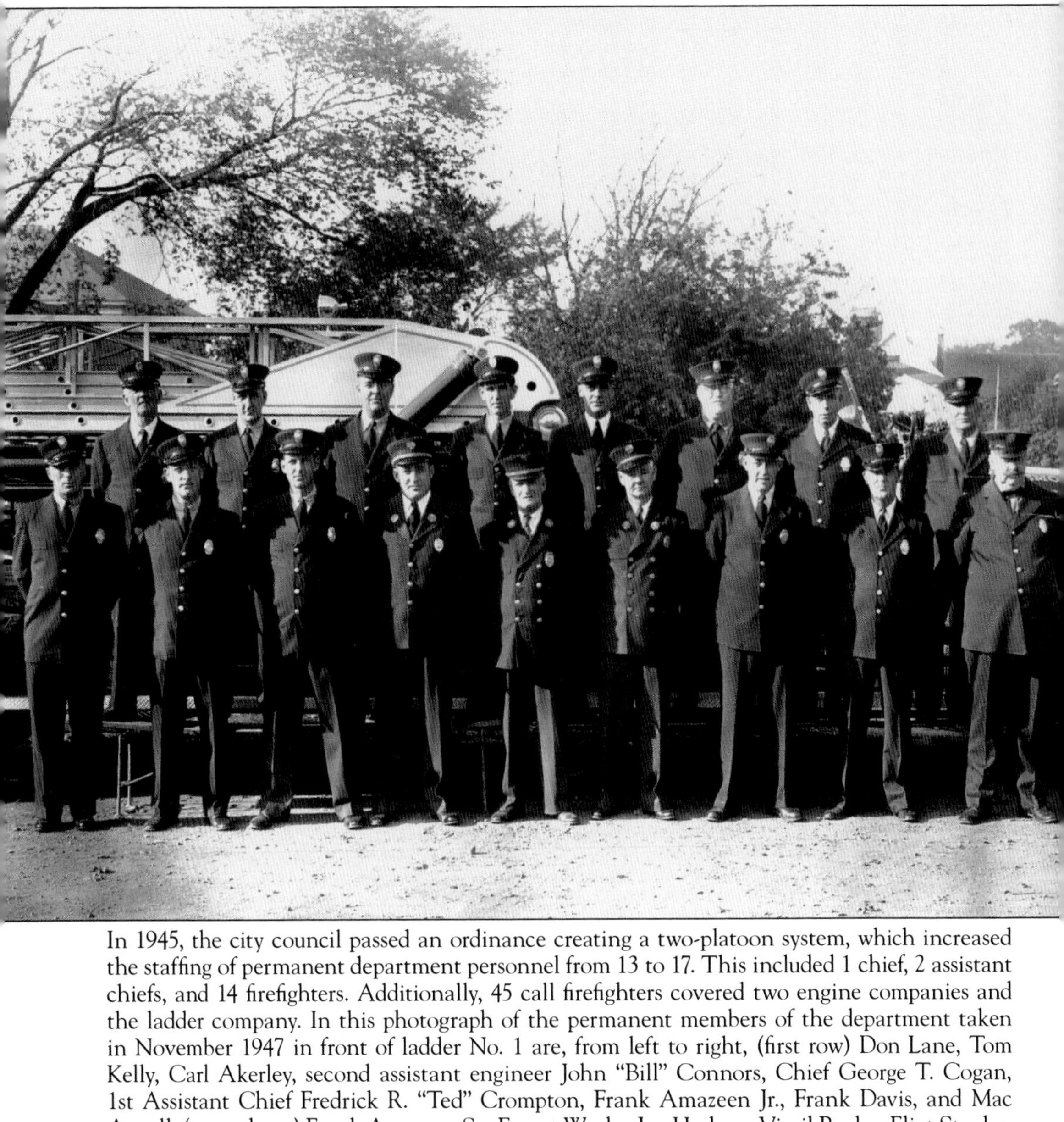

In 1945, the city council passed an ordinance creating a two-platoon system, which increased the staffing of permanent department personnel from 13 to 17. This included 1 chief, 2 assistant chiefs, and 14 firefighters. Additionally, 45 call firefighters covered two engine companies and the ladder company. In this photograph of the permanent members of the department taken in November 1947 in front of ladder No. 1 are, from left to right, (first row) Don Lane, Tom Kelly, Carl Akerley, second assistant engineer John "Bill" Connors, Chief George T. Cogan, 1st Assistant Chief Fredrick R. "Ted" Crompton, Frank Amazeen Jr., Frank Davis, and Mac Atwell; (second row) Frank Amazeen Sr., Ernest Weeks, Joe Hudson, Virgil Bagley, Eliot Staples, Darby Cox, Arthur Patterson, and Duncan Cameron. (Courtesy of Margaret Crompton.)

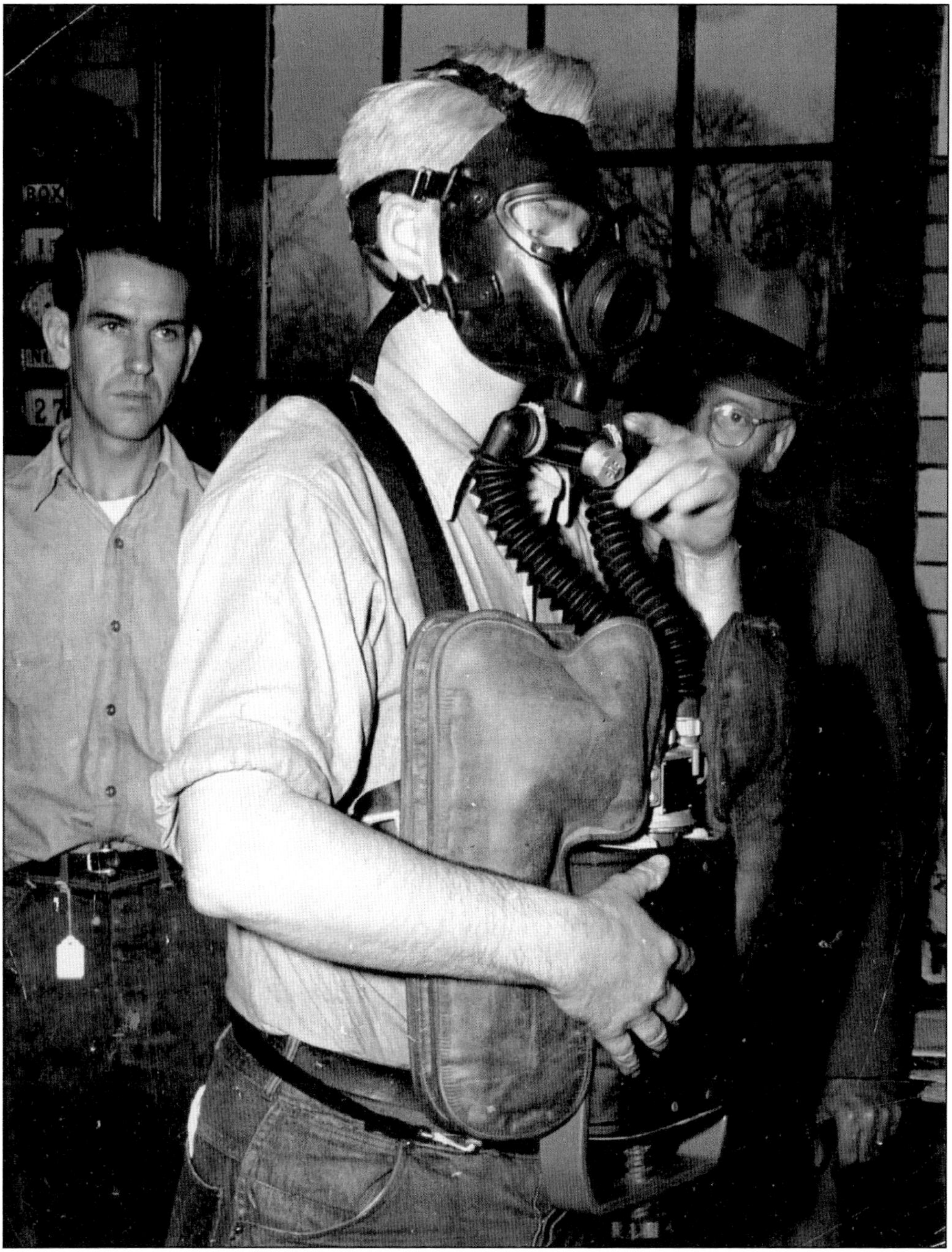

During the 1950s, the Portsmouth Fire Department experimented with early self-contained breathing apparatus (SCBA) known as rebreathers, which would recirculate and filter the wearer's own exhaled air. The units were found to be rather cumbersome and impractical for use. During the 1960s, the department would move toward the open-circuit SCBA of today. Here firefighter Virgil Bagley (left) and underwriter Ted Gunn (right) look on as firefighter Fredrick R. "Ted" Crompton (center) operates this new piece of equipment.

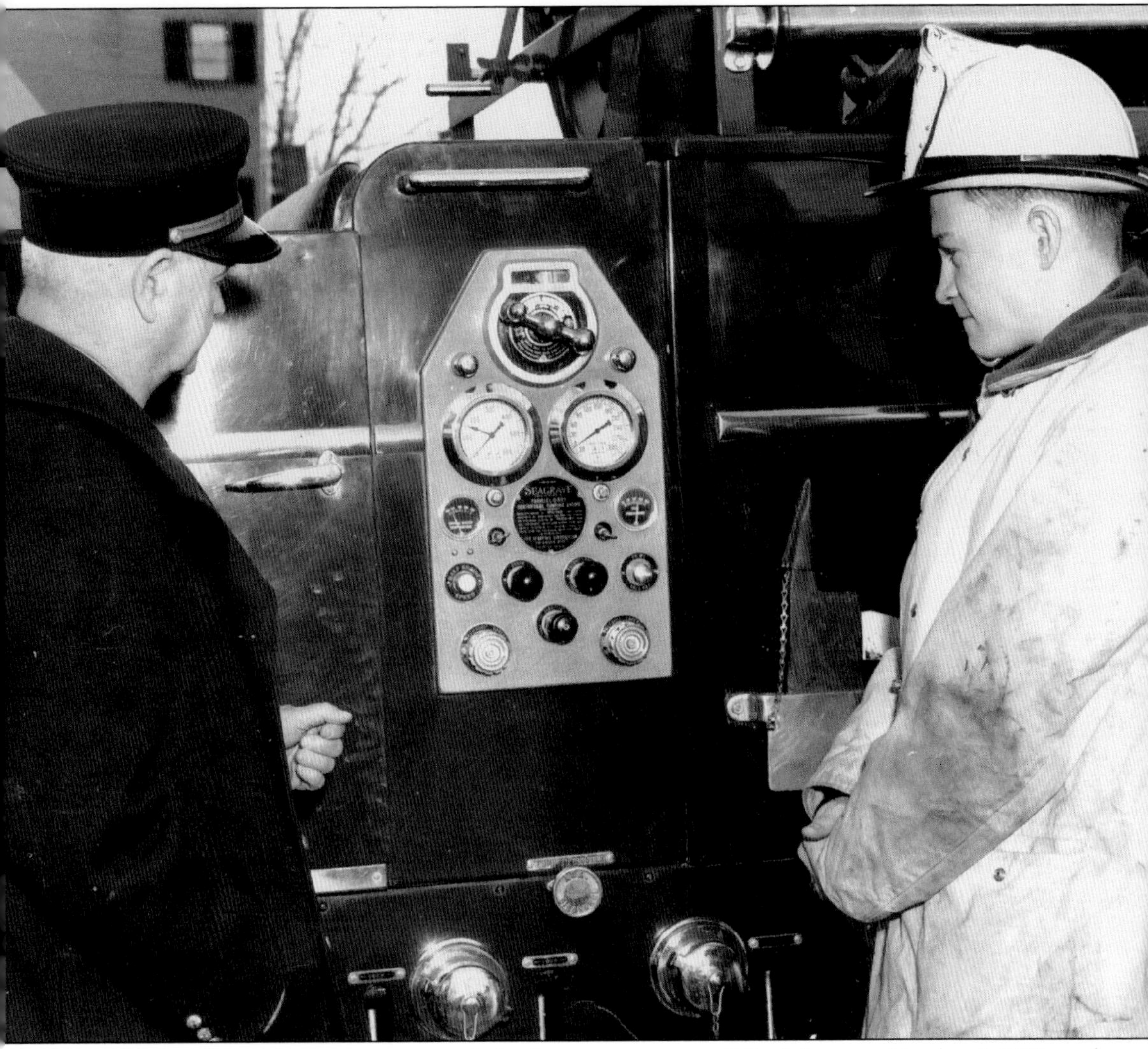

Fredrick R. "Ted" Crompton joined the department as a call firefighter on October 5, 1921, and was assigned to Chemical Engine No. 5. On April 4, 1927, he became a permanent firefighter. Crompton would rise through the ranks, serving as a senior captain, second assistant engineer, and first assistant engineer. On October 16, 1952, he would be appointed as the department's fire chief. Crompton retired on January 1, 1967. In this image taken on March 16, 1961, Chief Crompton (left) explains the operation of a fire pump to "fire chief for a day" David Wilson (right).

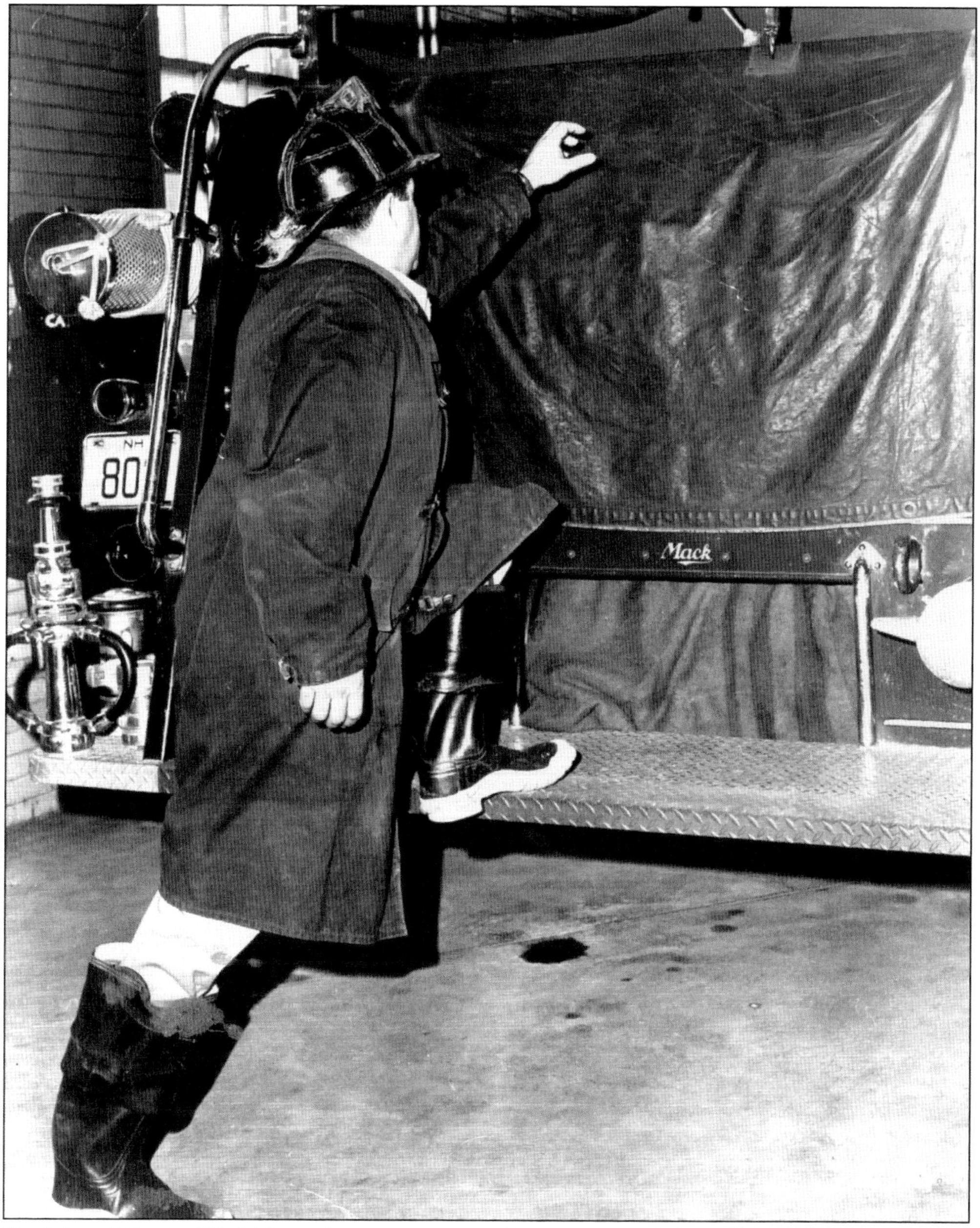

Firefighter Nicolo (Nick) Ingoglia jumps up on the rear step of the 1950 Mack engine at the Central Fire Station in this image from the mid-1960s. Firefighter Ingoglia joined the department as a call man in 1953 and was appointed as a permanent firefighter on June 1, 1957. Riding the tailboard was a common practice for firefighters responding to and returning from calls until the early 1980s.

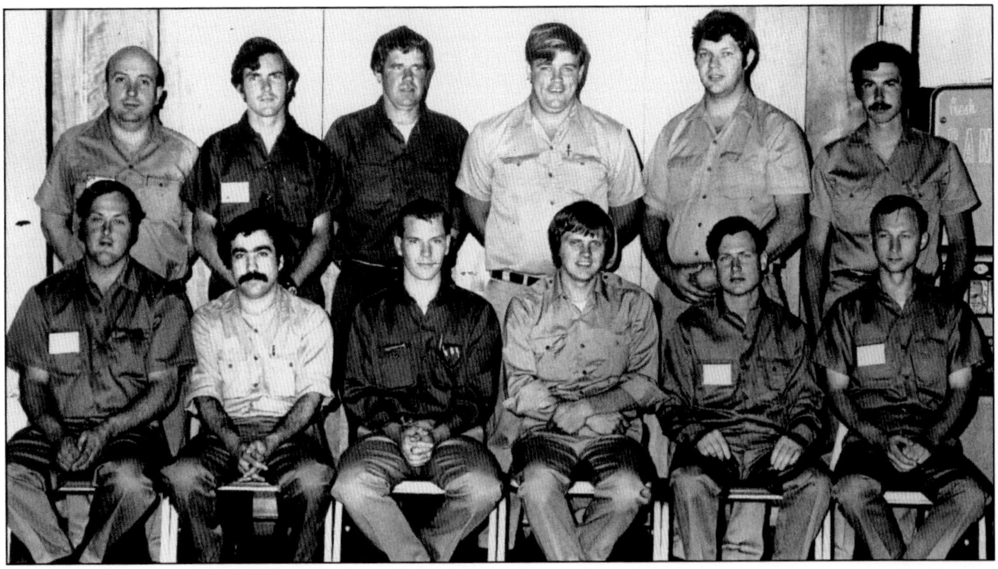

In 1974, the hours worked by Portsmouth firefighters were reduced from 63 to 48. As a result, 12 new firefighters were added to the force. These recruits were introduced to the fire service through three tough weeks at the first-ever New Hampshire Firefighter Recruit School. In this photograph taken on June 10 of that year are, from left to right, (first row) Clifton Boone, Andrew Dipietro, Herald Good, James Coughenour, Frank Ott, and Joseph Snook; (second row) Walter Wakefield, Richard B. Duddy, Ralph Hammer, Thomas Farmham, George Francois III, and Gerard Goyette.

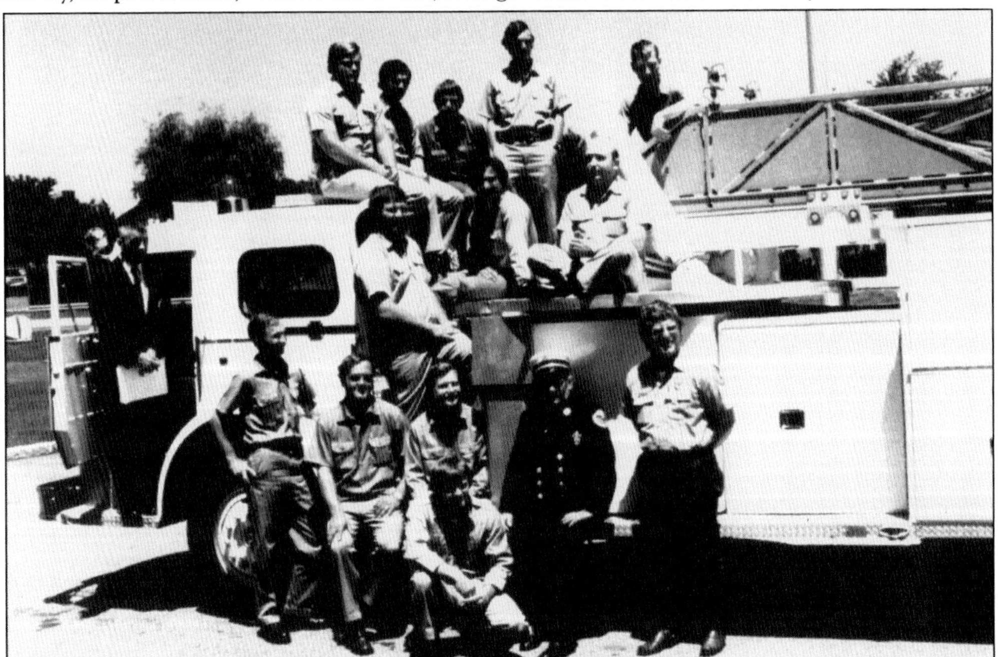

Known in Portsmouth Fire Department lore as the "dirty dozen," Portsmouth Fire Department recruits along with their instructors pose for a photograph during the 1974 New Hampshire Firefighter Recruit School. The training was held at the New Hampshire Technical College located at the corner of the Route 1 bypass and Greenleaf Road in Portsmouth.

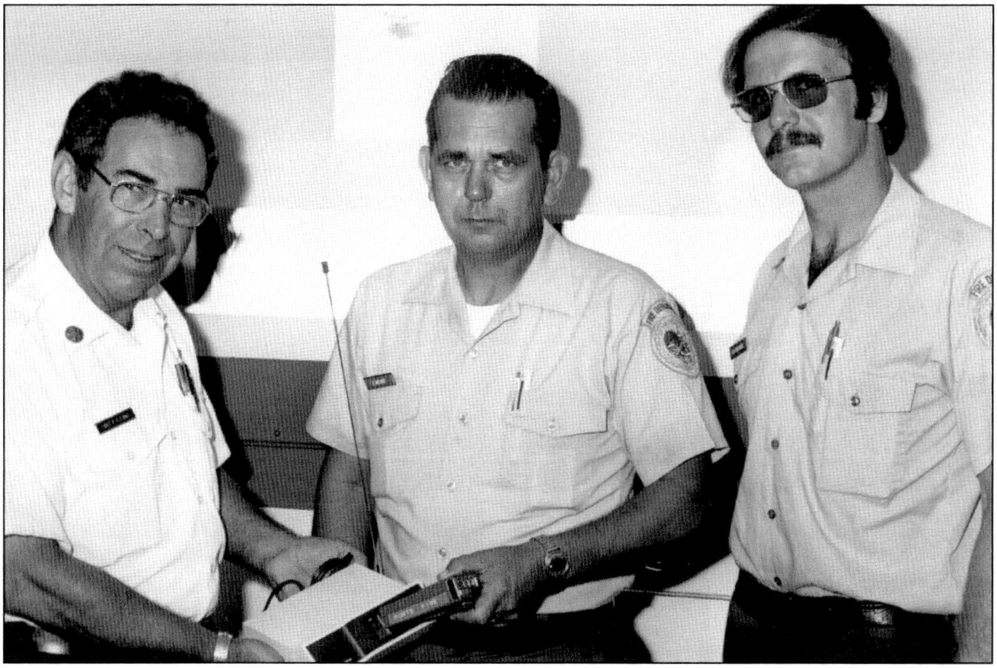

In this August 1977 photograph are, from left to right, Chief Paul Long, firefighter Everett "Al" Weare, and firefighter Theodore (Ted) Boisvert as they pose with a new Motorola Minitor pager that included an audio amplifier-charger. Prior to the purchase of these pagers, department members were recalled for fires individually by telephone. For a cost of around $350, these portable radio scanners were small enough to be worn by firefighters and only activated when needed.

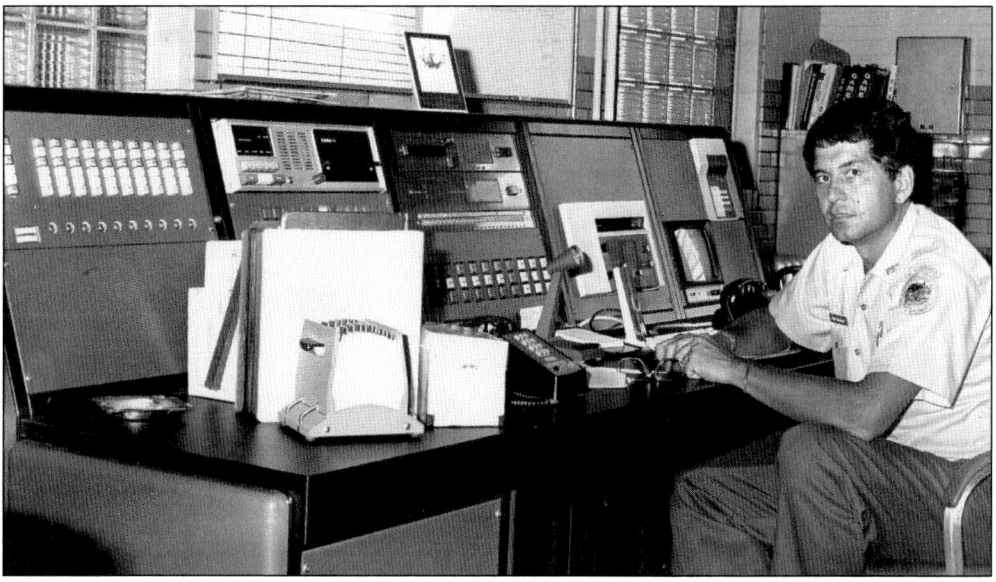

Firefighter Gene Franzoso is seen here in 1978 during his assigned "desk watch." The watch room was manned by on-duty firefighters at the Central Fire Station on Court Street. In 1957, the new watch room was tied into a state-of-the art fire alarm system, which was connected to 83 fire alarm boxes. By 1993, fire-dispatching duties were transferred to a combined fire and police central dispatch center at the Portsmouth Police Station on Junkins Avenue.

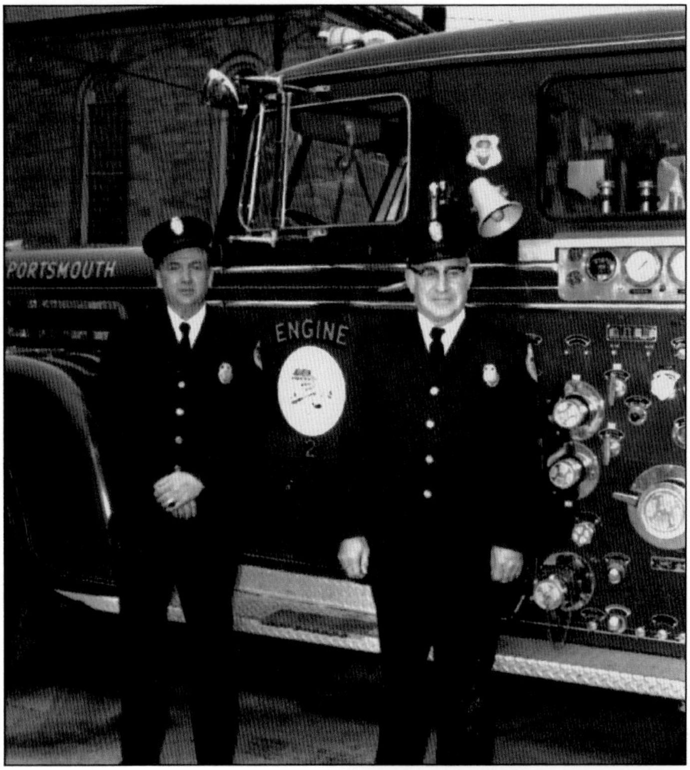

Portsmouth firefighters Carl Widen (left) and John "Bill" Connors (right) are seen here on February 2, 1977, next to combination No. 1. Firefighter Widen was a call man assigned to Combination No. 2 beginning in 1955. He was appointed as a permanent member of the department on April 1, 1959. Both he and firefighter Connors retired from the department in 1976—firefighter Widen in December and firefighter Connors in October.

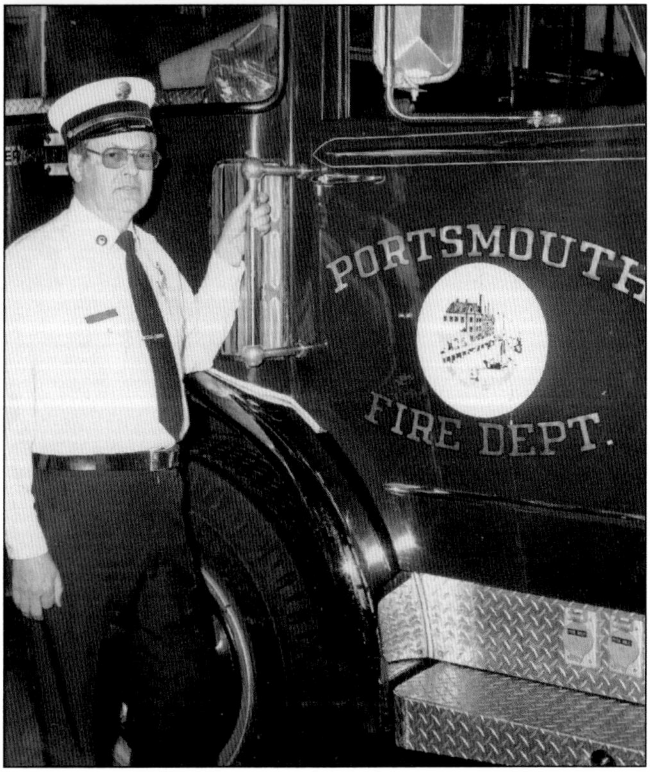

On March 1, 1979, Deputy Chief Eugene Hersey retired after 26 years of service with the department. He joined the department as a call firefighter in April 1953 and within a month was appointed to a permanent position. Throughout his career, firefighter Hersey would serve as a lieutenant starting in 1967, captain in 1969, and deputy chief beginning in 1974. He also served a number of years as the department's training officer.

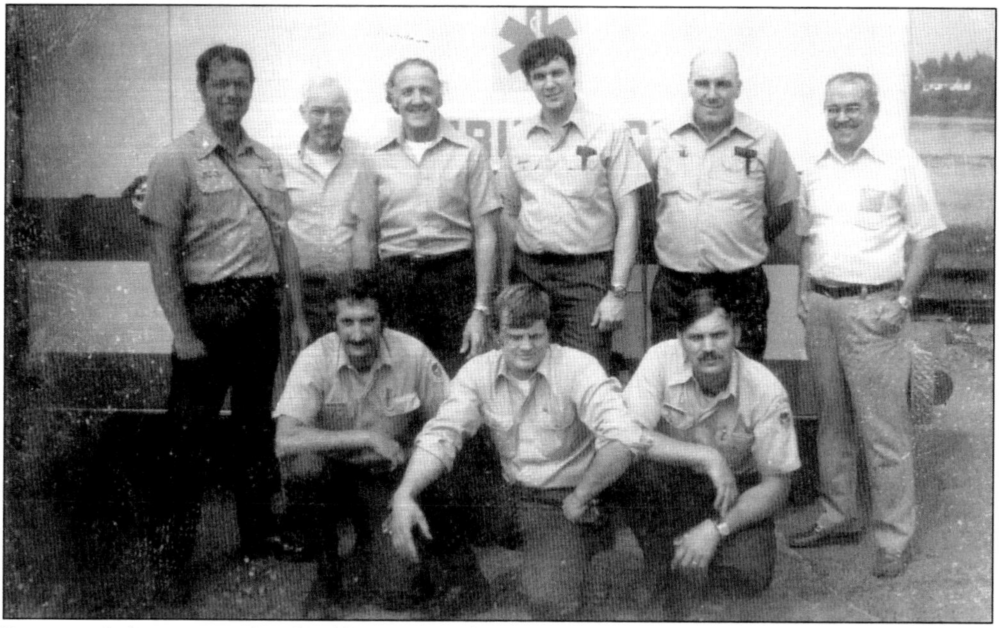

Portsmouth fire personnel and Public Service of New Hampshire (PSNH) workers pose alongside the ambulance in this July 1979 photograph. The department had just participated in a tour of PSNH's Schiller Power Station on Gosling Road. Seen here are, from left to right, (first row, kneeling) Dan Hovey, Jim Coughenour, and Stan Whitney; (second row) Robin Hughes, two unidentified PSNH employees, Steven Griswold, George Ross, and another unidentified PSNH employee.

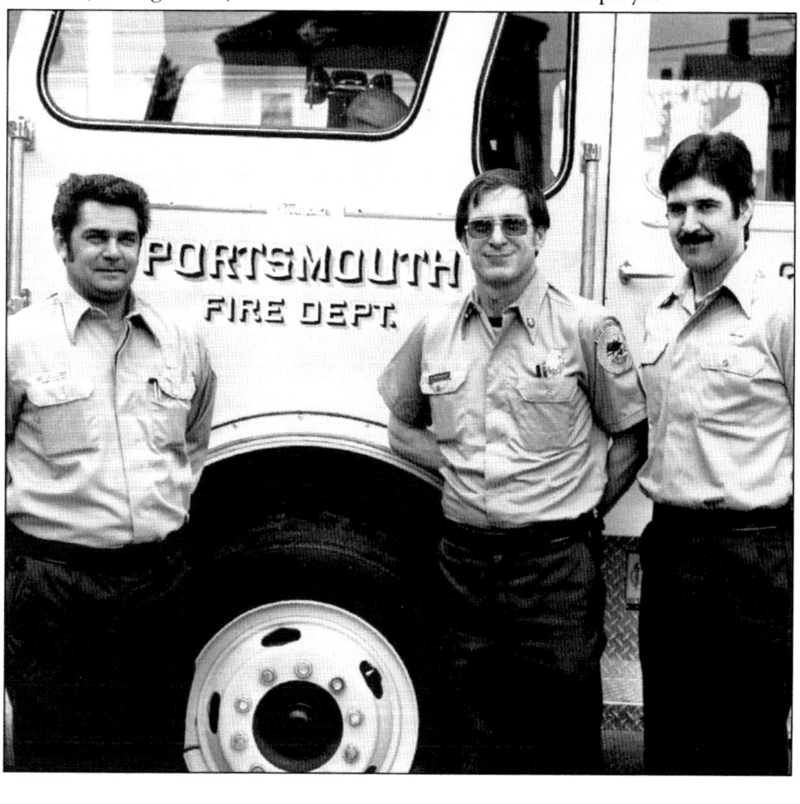

In this photograph taken in front of Portsmouth's Maxim pumper around 1981 are, from left to right, firefighter Art Fortin, Capt. Ralph DiBenardo, and firefighter Val Pamboukes. This pumper was only one of two department fire engines ever to be painted lime green. The other during the same period was a Mack fire engine.

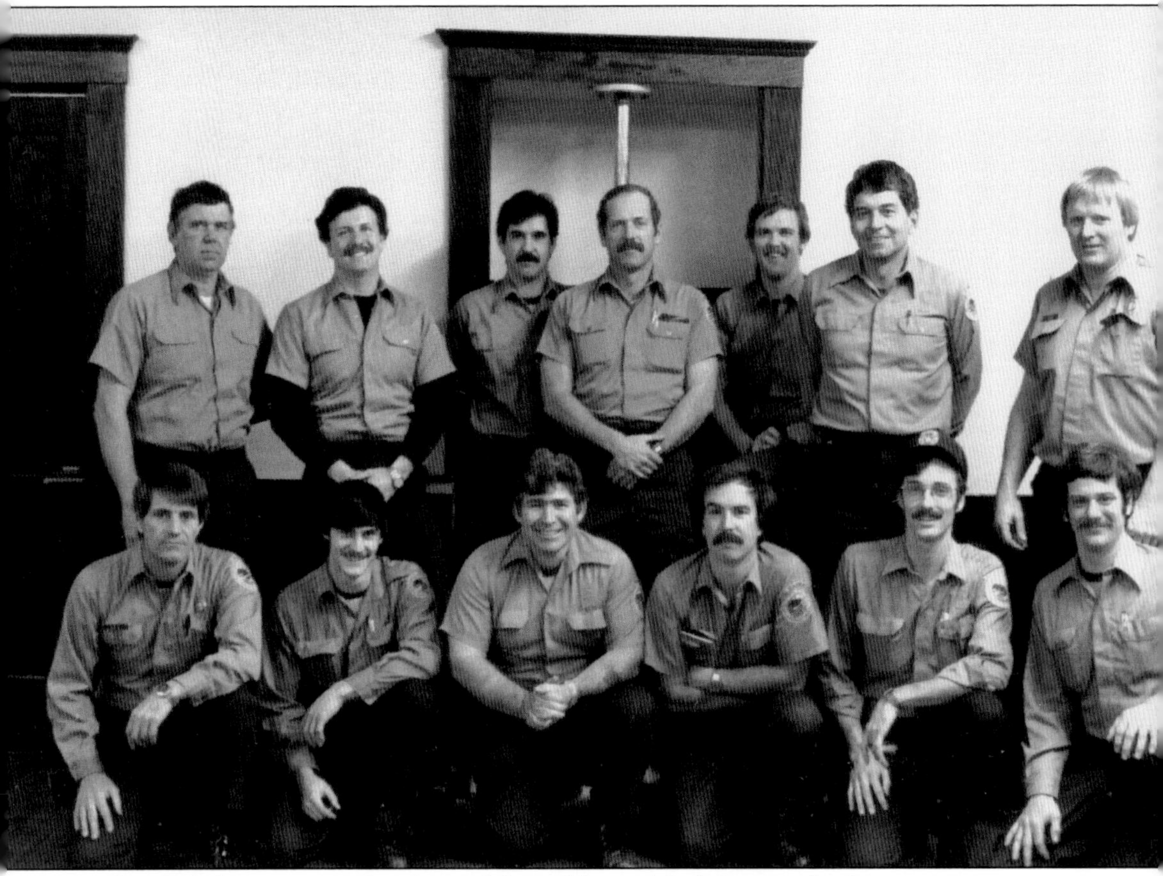

In 1986, the city council authorized four additional firefighters and a fire prevention person to be added to the department, bringing the organization to a total of 60 firefighting personnel. Some 20 firefighters were certified emergency medical technicians (EMT), and all others were Red Cross Advanced First Aid qualified with additional training in patient assessment, oxygen therapy, and vehicle extrication. All newly assigned firefighters were trained to the EMT level. There were four duty shifts, and each had a complement of 13 personnel. Pictured here are, from left to right, the members of D-Shift next to the fire pole on the second floor at the Central Fire Station: (first row) Capt. Edward Tulley, David "Pat" Sliney, Edward Daneault, Gerard Goyette, Paul Trivingo, and James Rivais; (second row) Ralph Hammer, Ted Boisvert, Vassilios "Val" Pamboukes, James Wooley, Charles Jones, Gene Franzoso, and Lt. Harold "Joe" Dow.

Four

Answering the Call

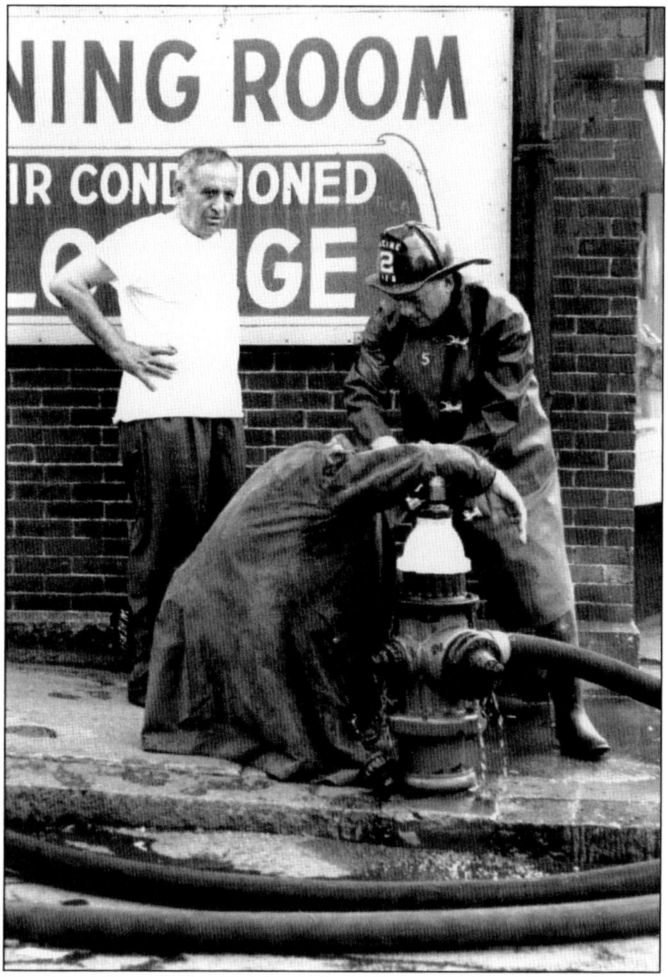

John Paola, in the white T-shirt, looks on as an unidentified firefighter from Engine No. 2 comes to the aid of firefighter William J. Hersey as he collapses from exhaustion against a fire hydrant at Deer and Vaughn Streets in this photograph taken sometime in the late 1950s. Firefighter Hersey was a call man assigned to Engine No. 2 beginning on July 1, 1957.

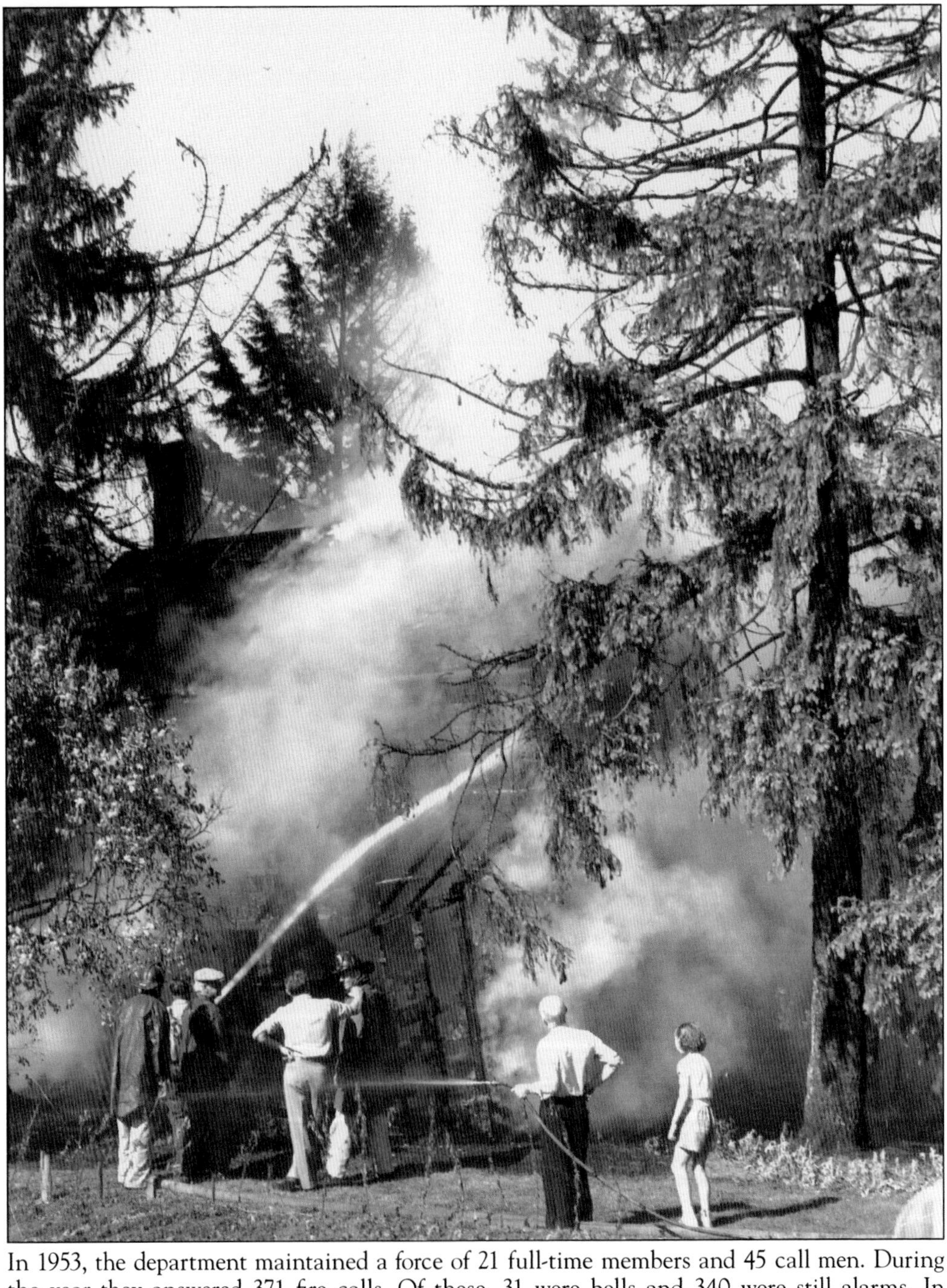

In 1953, the department maintained a force of 21 full-time members and 45 call men. During the year they answered 371 fire calls. Of these, 31 were bells and 340 were still alarms. In addition, the department's ambulance responded to 695 calls. Property damage caused by fire during the year amounted to $104,772. In this photograph taken on the afternoon of June 3, 1953, the neighborhood came out for this fire that ignited in a two-story vacant barn owned by J. Dinnerman between Wibird and Union Streets. It took over an hour for Portsmouth firefighters to extinguish the blaze, despite the assistance of a well-intentioned "firefighter" and his garden hose.

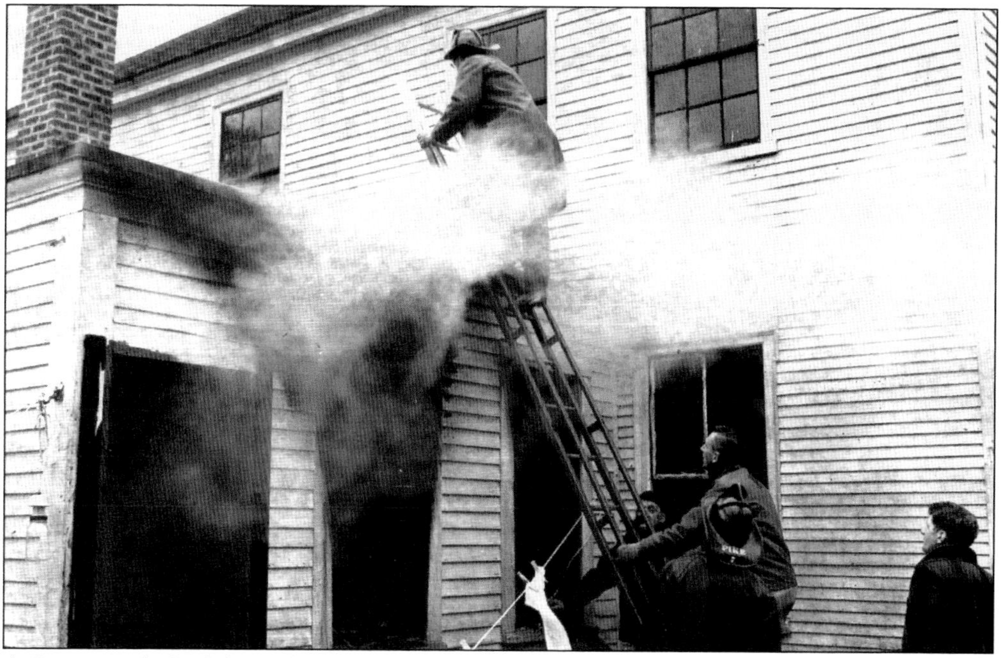

A Portsmouth firefighter checks for fire extension at this 200 Marcy Street duplex fire on December 6, 1954. A mother and her two-year-old daughter escaped the thick black smoke and found refuge from the winter chill in a neighbor's apartment. Both waited close to an hour while Portsmouth firefighters extinguished and overhauled the charred interior. (Courtesy of the Portsmouth Herald.)

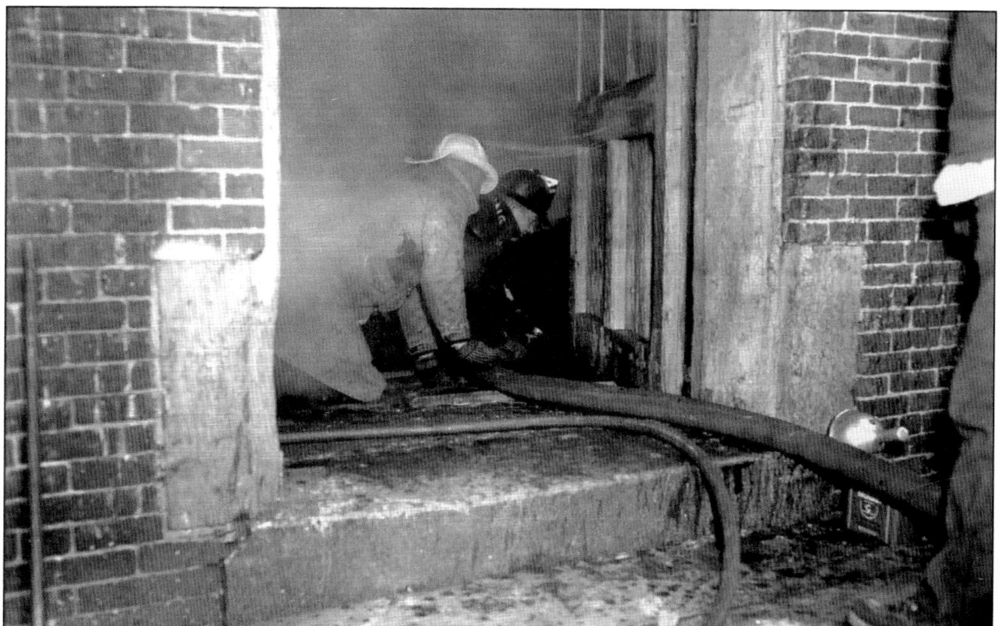

Portsmouth firemen were hampered by freezing temperatures in battling a difficult fire at the launderette at 690 Lafayette Road on January 9, 1959. Flames shot from the basement into the area of the clothes dryers and heavily damaged this brick building erected in 1883.

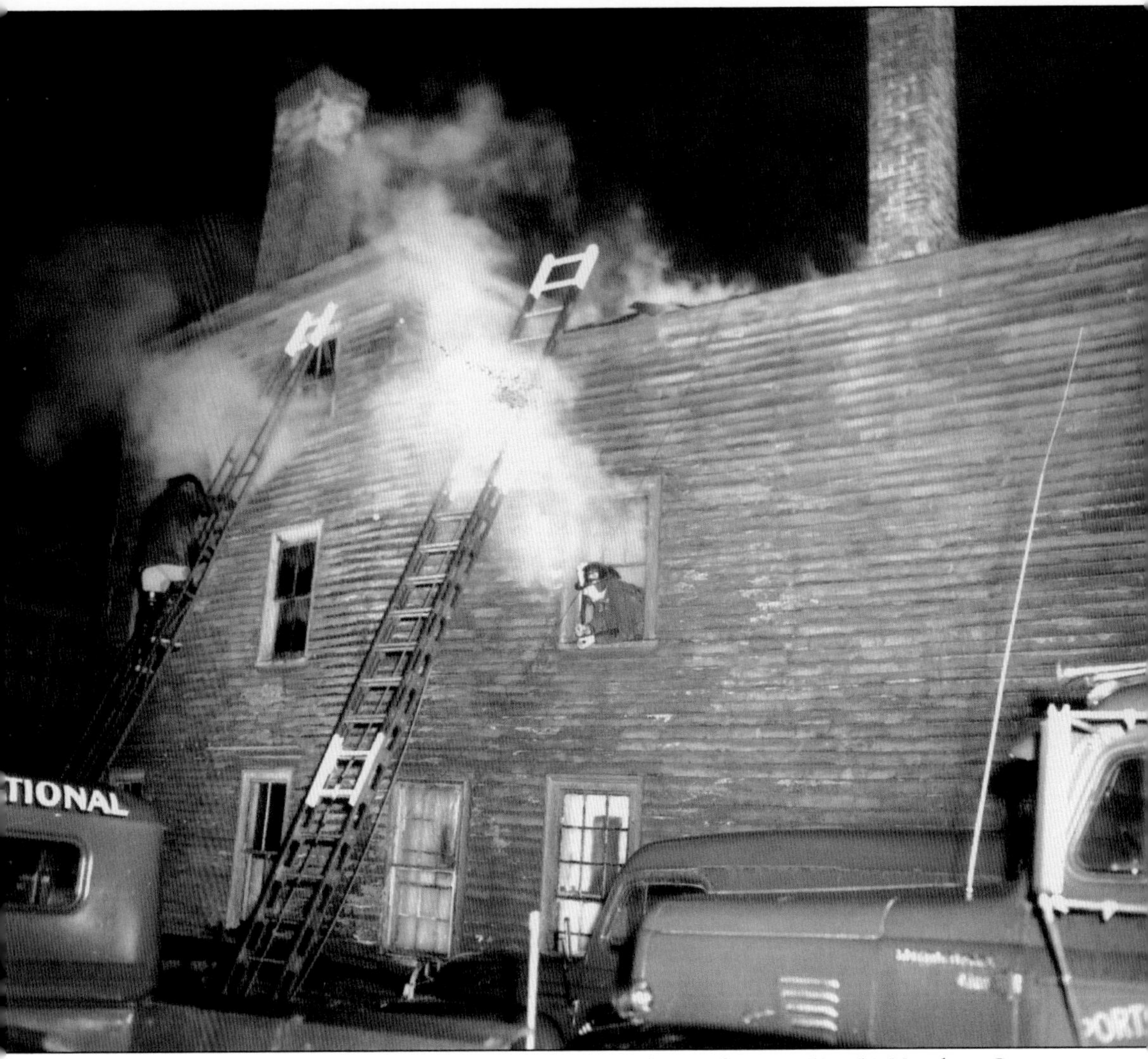

A flooded oil burner touched off this apartment house fire at 152–154 Vaughan Street on February 4, 1959. Firefighters arrived at 7:18 p.m. and were released two hours later when the all clear was sounded. Extensive smoke and water damage occurred throughout the structure. Multiple ground ladders are positioned, and an unidentified Portsmouth firefighter gets some fresh air in this photograph taken during fire operations. In 1959, the department answered 1,117 total calls, which included 363 still alarms, 59 bells, and 661 ambulance responses. An additional 11 calls were made to the New Hampshire towns of Rye, Greenland, Newcastle, and Newington and Eliot, Maine, under the mutual aid system.

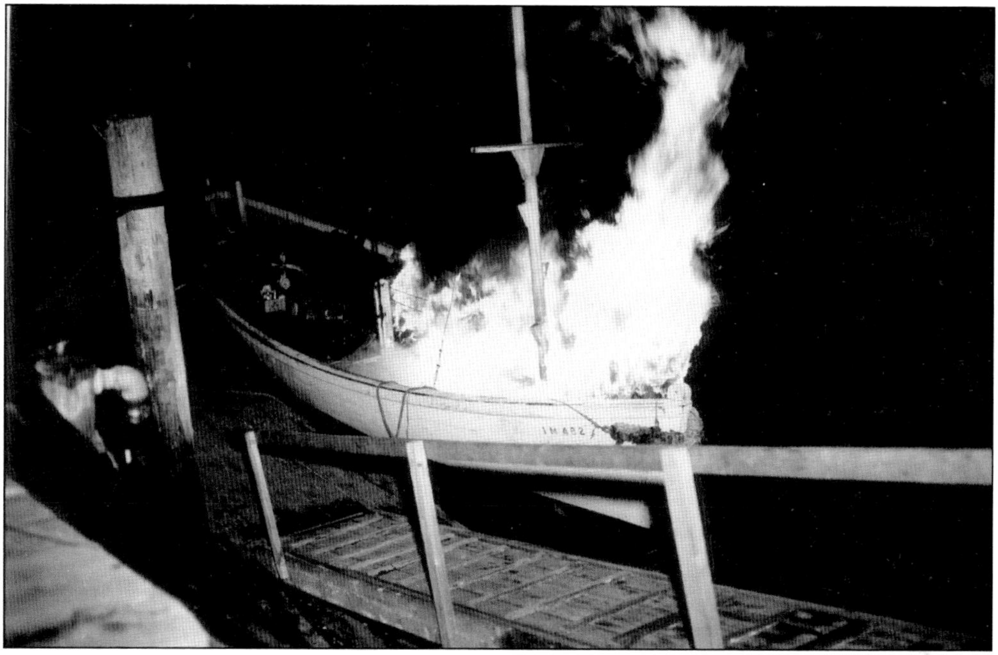

A July 28, 1959, fire that started in the engine bulkhead of a fishing boat along the city's waterfront sent two Maine men to the hospital with burns. Firefighters responded to box 24 at the corner of Bow and Daniel Streets at 11:30 p.m. on July 28, 1959, to find the vessel ablaze at a wharf of the Portsmouth Fish and Lobster Company. Shortly after their arrival, an explosion occurred, sending a fireball into the night sky. It was reported that the main fuel tank along with two or three additional smaller cans of gasoline contributed to the massive blast. It took fire crews more than an hour before the fire was reported out.

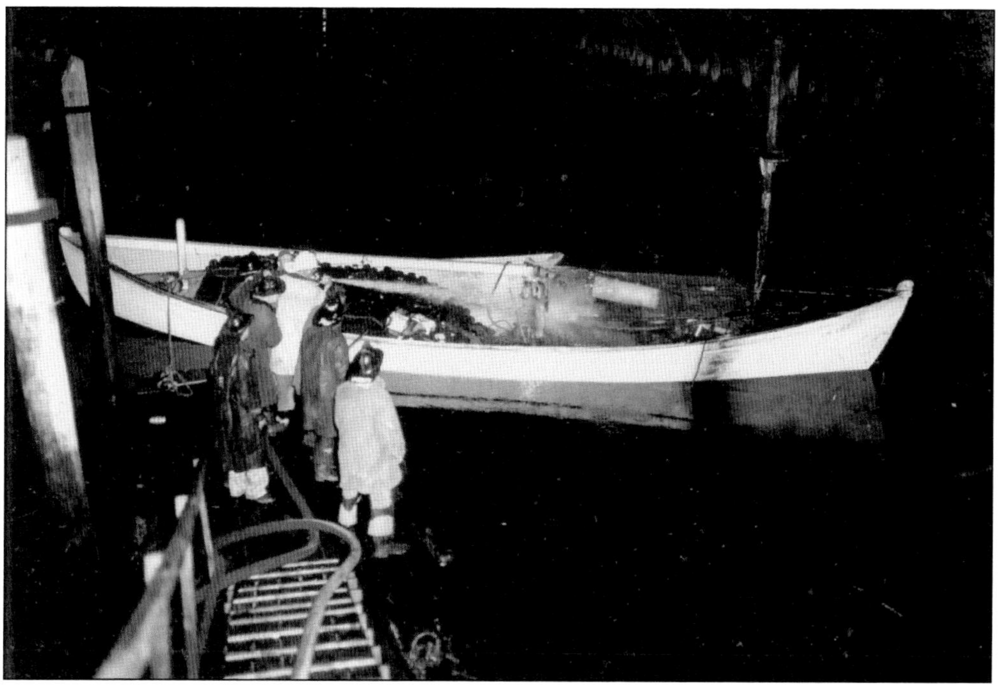

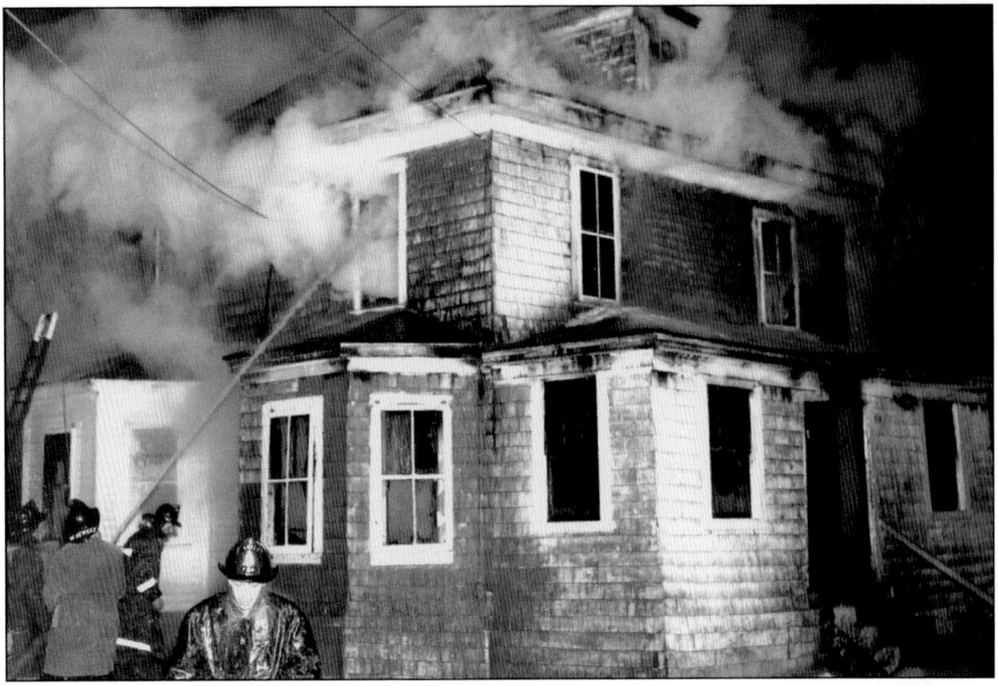

Three residents had to flee into the subzero temperatures during an early-morning blaze on Versailles Avenue that was believed to have been caused by an overheated furnace. According to Fire Chief Fredrick R. "Ted" Crompton, the February 2, 1961, fire started around the chimney and quickly engulfed the first floor. Portsmouth firefighters direct hose lines through a second-floor window as smoke rolls out the upper floors and eves in this image taken during the fire.

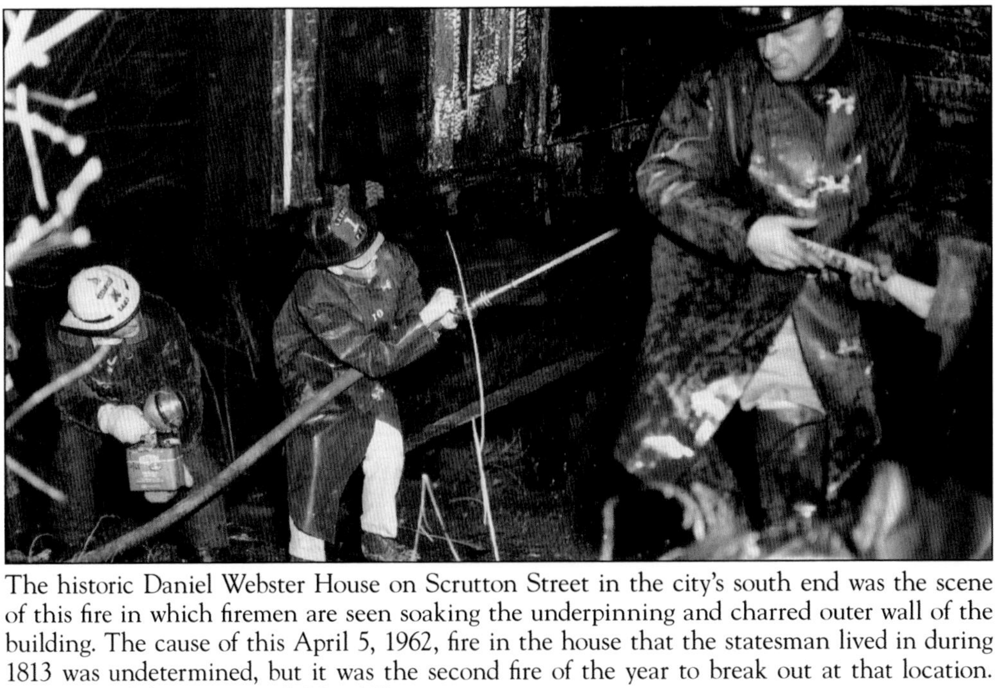

The historic Daniel Webster House on Scrutton Street in the city's south end was the scene of this fire in which firemen are seen soaking the underpinning and charred outer wall of the building. The cause of this April 5, 1962, fire in the house that the statesman lived in during 1813 was undetermined, but it was the second fire of the year to break out at that location. (Courtesy of the Portsmouth Herald.)

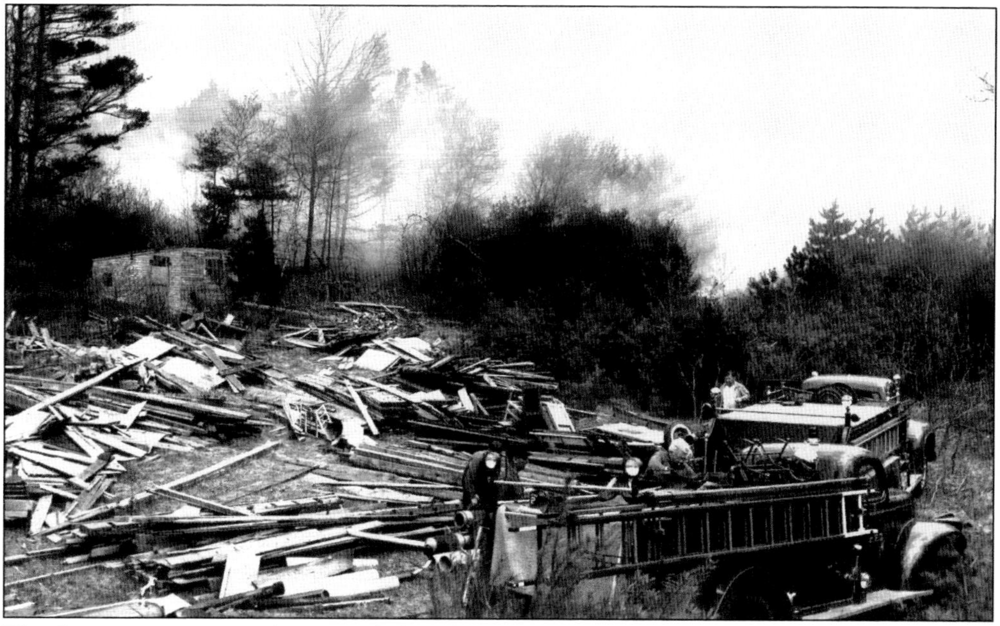

The department constantly responded to the city dump off Jones Avenue for debris and wood fires during the 1950s and 1960s. Most of the fires at the dump were caused by burning embers from the city's incinerator that touched off small brush fires. On October 4, 1962, lumber from houses that were torn down and the surrounding woods caught fire, requiring a multiple-alarm response by the department. Fire companies worked for over four hours with booster lines, portable five-gallon water pumps, and forestry hand tools in order to control the blaze.

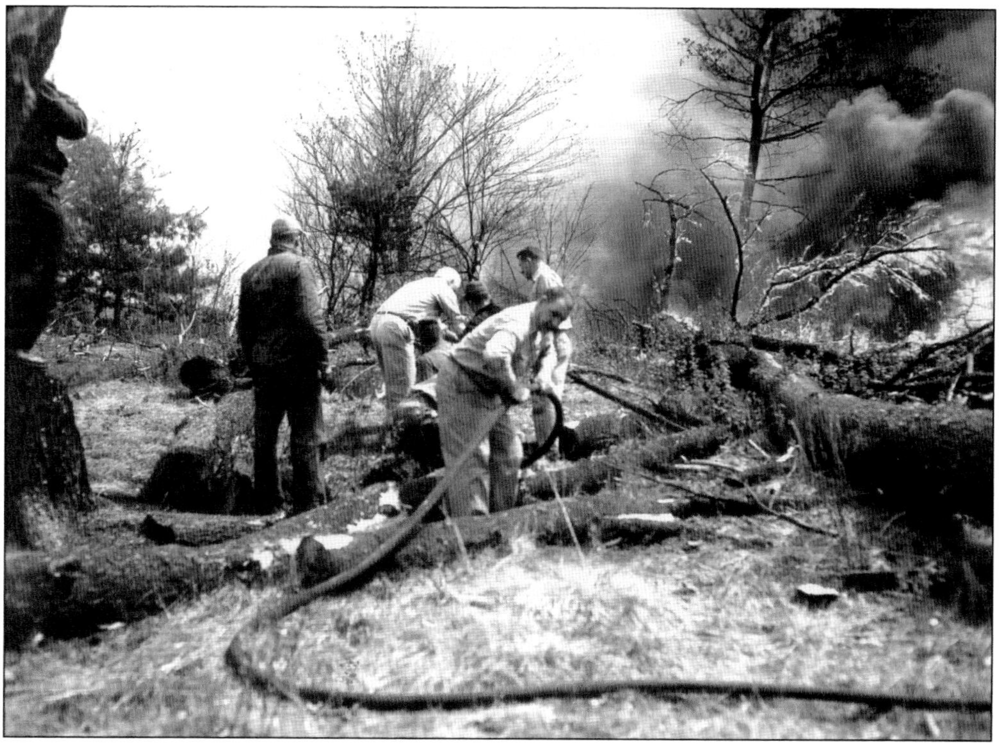

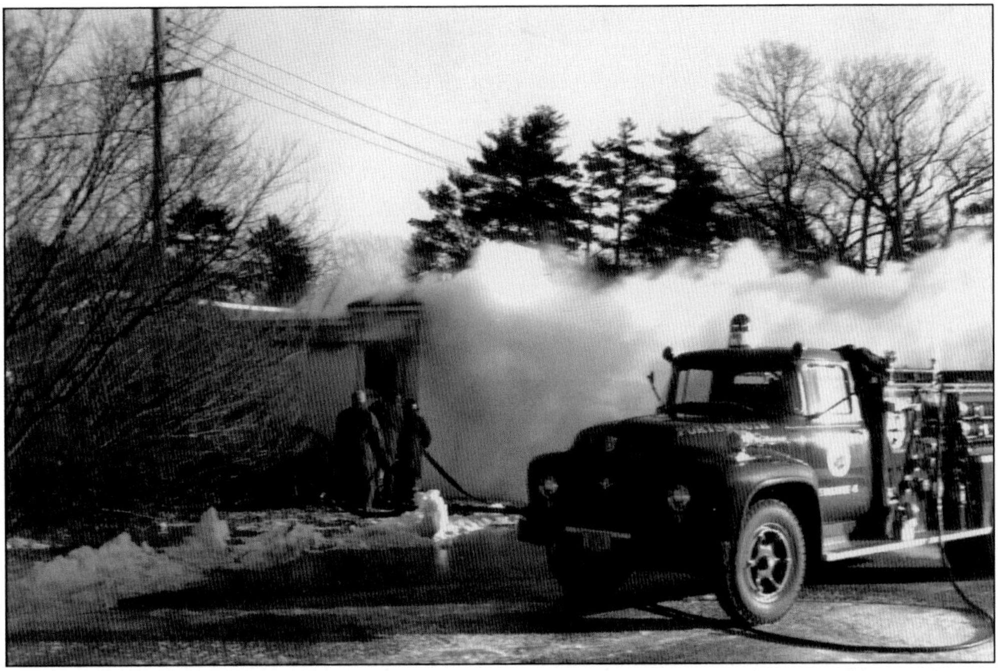

Firefighters faced icy winds during an early-morning January 24, 1963, fire at Ladd's Ice Cream on Sagamore Avenue. Dense smoke billowed from the building for close to an hour during the stubborn blaze, while cans of ice cream created numerous small explosions. Flames and sparks also emanated from a transformer at the rear of the structure, hampering firefighting efforts until the power could be shut off by the utility company. (Courtesy of the Portsmouth Herald.)

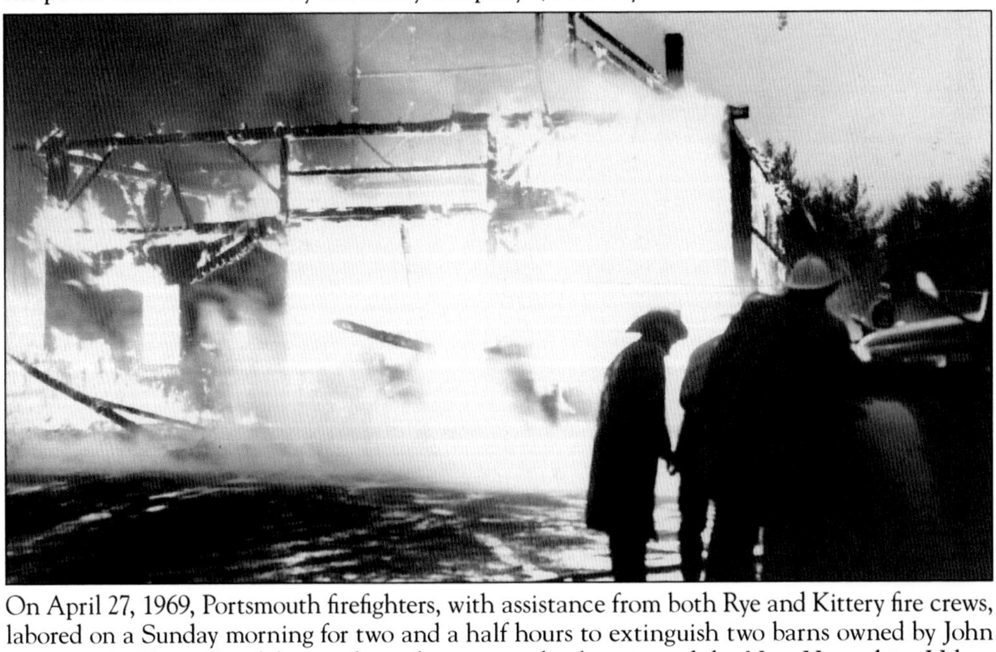

On April 27, 1969, Portsmouth firefighters, with assistance from both Rye and Kittery fire crews, labored on a Sunday morning for two and a half hours to extinguish two barns owned by John Stone at 50 Elwyn Road (across from the present-day location of the New Hampshire Urban Forestry Center). The barn here, built with handwrought nails and wooden pegs, was completely destroyed. (Courtesy of the Portsmouth Herald.)

Smoke pours from this two-story dwelling at 247 Cass Street on December 12, 1976. Firefighters battled intense heat and thick smoke in order to control this structure fire that broke out shortly after 9:00 a.m. The kitchen and pantry areas on the first floor were totally destroyed while substantial heat and smoke damage were evident throughout the building (right). Firefighter Harold "Buster" Sullivan is seen with a cat rescued from the home during the fire (below).

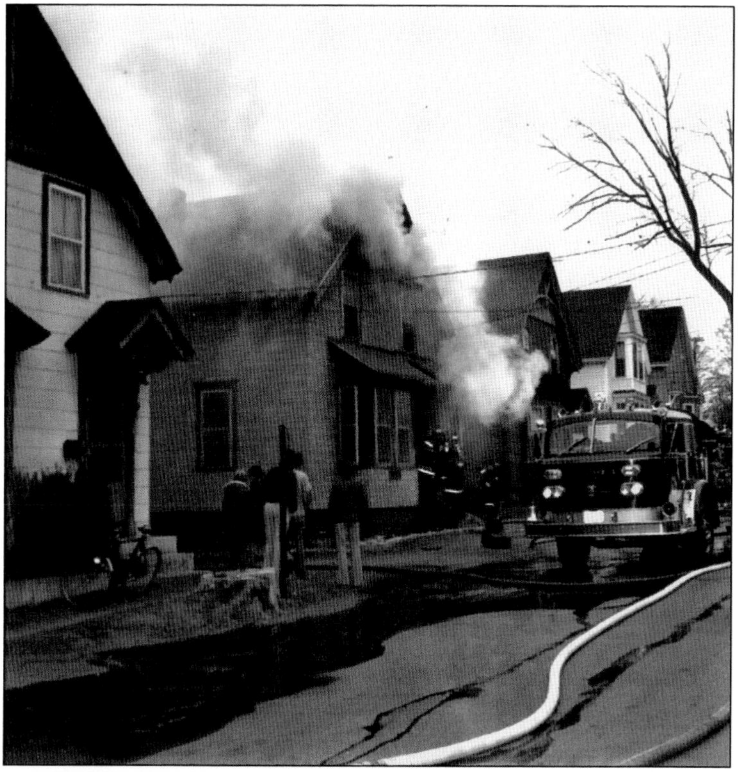

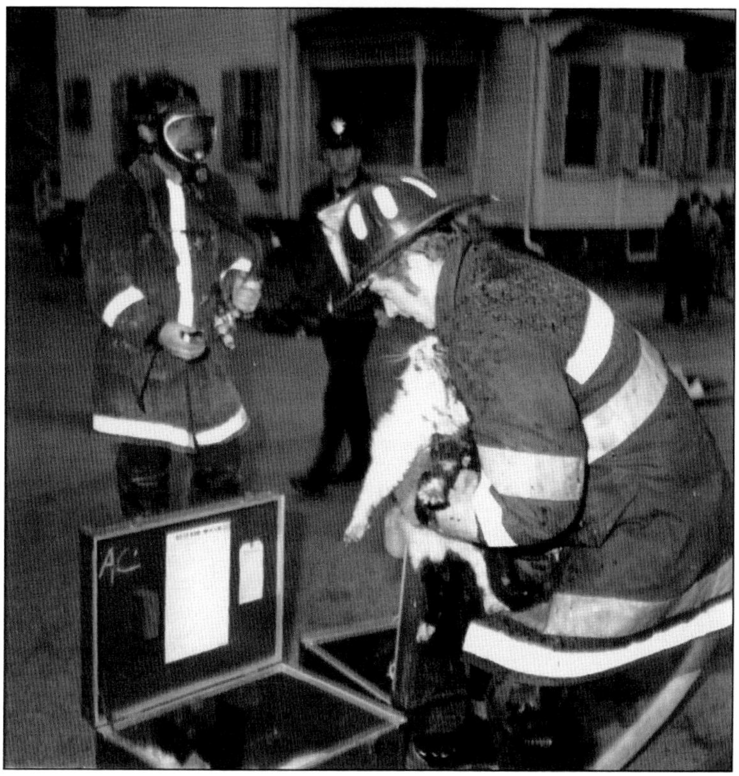

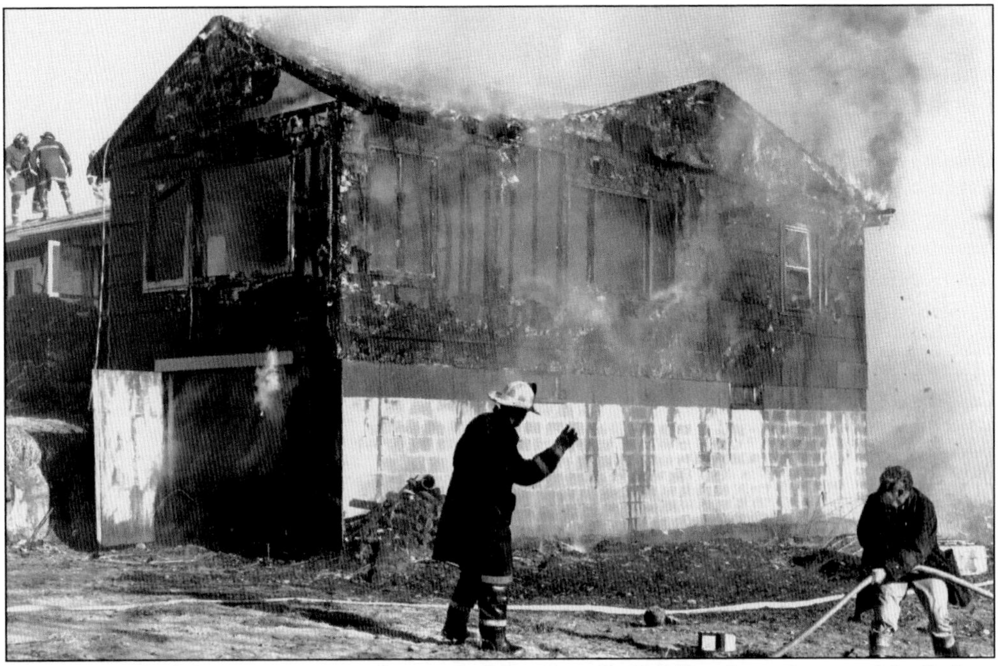

Portsmouth firefighters operate on the roof of this Peverly Hill Road ranch owned by Robert Dowd while Deputy Chief Eugene (Gene) Hersey assesses the damage to the garage caused by this mid-morning blaze that occurred on November 11, 1976. Although the home suffered considerable damage, no injuries were reported.

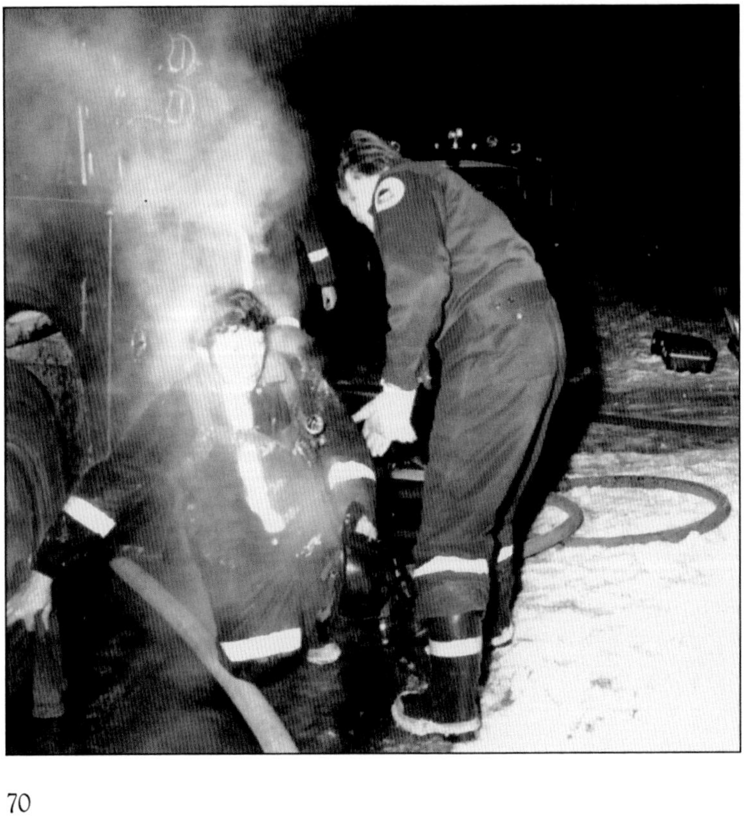

Three city firefighters were treated at Portsmouth Regional Hospital for smoke inhalation and exhaustion as a result of fighting a basement fire in a vacant Maple Haven ranch late on Christmas Eve in 1976. Portsmouth firefighter George "Butch" Francois (left) braces himself against one of the department's engines as he receives assistance from firefighter Stanley Whitney (right) in this photograph taken during the fire. (Courtesy of the Portsmouth Herald.)

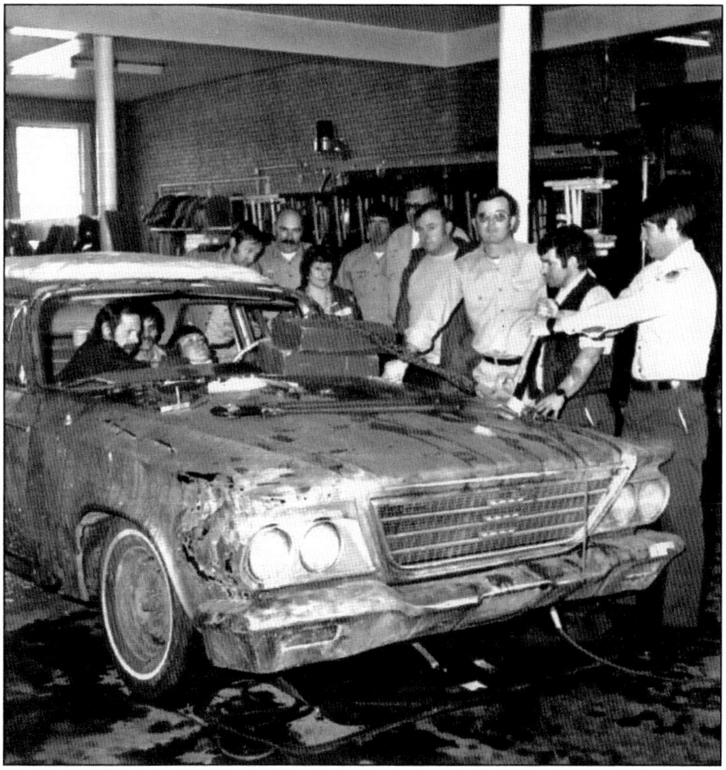

Portsmouth firefighters use a donated car as they practice treating a patient injured in a car crash during EMT training. They also capitalize on the warmth and comfort of the station as this image was taken inside the Court Street station on January 1, 1977. During 1977, the department would respond to 788 fire and 1,549 ambulance calls.

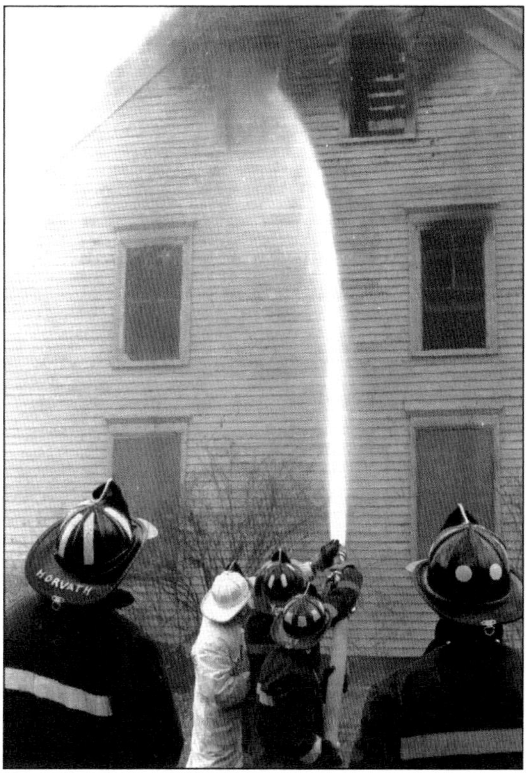

Portsmouth firefighters control a fire set to a Cate Street house as part of a mutual aid training exercise with the Pease Air Force Base Fire Department in this October 27, 1977, photograph. Seen here directing a two-and-a-half-inch hose line to a third-floor window in the center of the image are, from left to right, Deputy Chief Gene Hersey, Emery York, and Frank Ott as Baylor Horvath (far left) and another unidentified Portsmouth firefighter (far right) look on.

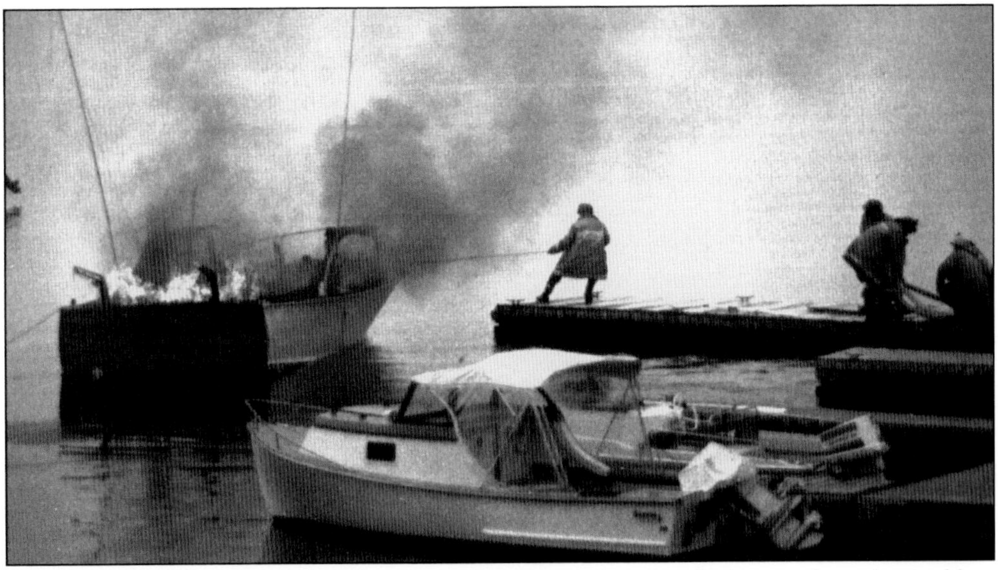

An unidentified Portsmouth firefighter struggles to maintain control of a powerboat that is ablaze and threatening other boats in this striking image taken around 1983. Additional crews can be seen advancing hose lines down the dock to extinguish the fire. The incident occurred at Mike's Marina, today known as Witches Cove Marina, off Wentworth Road.

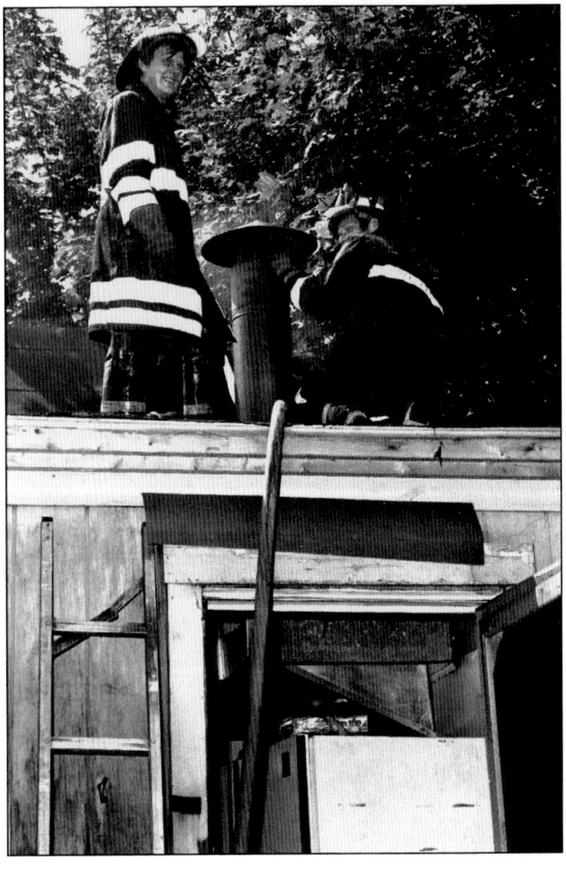

Portsmouth firefighters James E. Coughenour (left) and James B. Wooley (right) have a booster hose line stretched during overhaul operations on the roof at this August 27, 1979, fire that broke out at a local pub known as Duff's Dump on Islington Street. Both firefighter Coughenour and firefighter Wooley joined the department in 1974 and were appointed as permanent members in 1975.

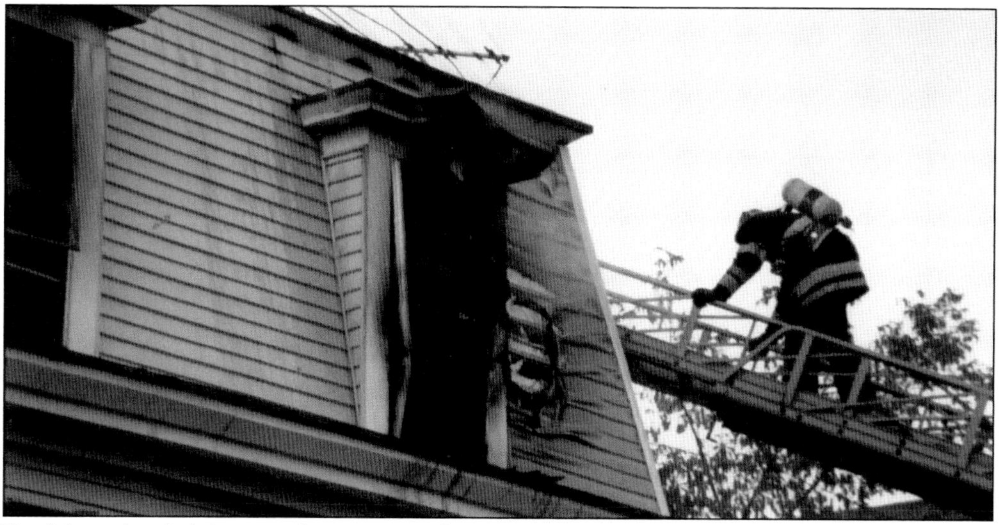

Firefighters battled the same fire twice at a four-unit apartment building at the corner of Islington and Dover Streets on Sunday, September 29, 1986. The first fire was fought in the cellar and was extinguished within an hour or so. Some 20 minutes after all units cleared, the department was summoned back to find the fire had crept inside the walls and into the attic. The fire heavily damaged the structure and left two residents homeless. In this photograph, an unidentified Portsmouth firefighter is seen working his way up the aerial of ladder No. 1 as the second fire is brought under control.

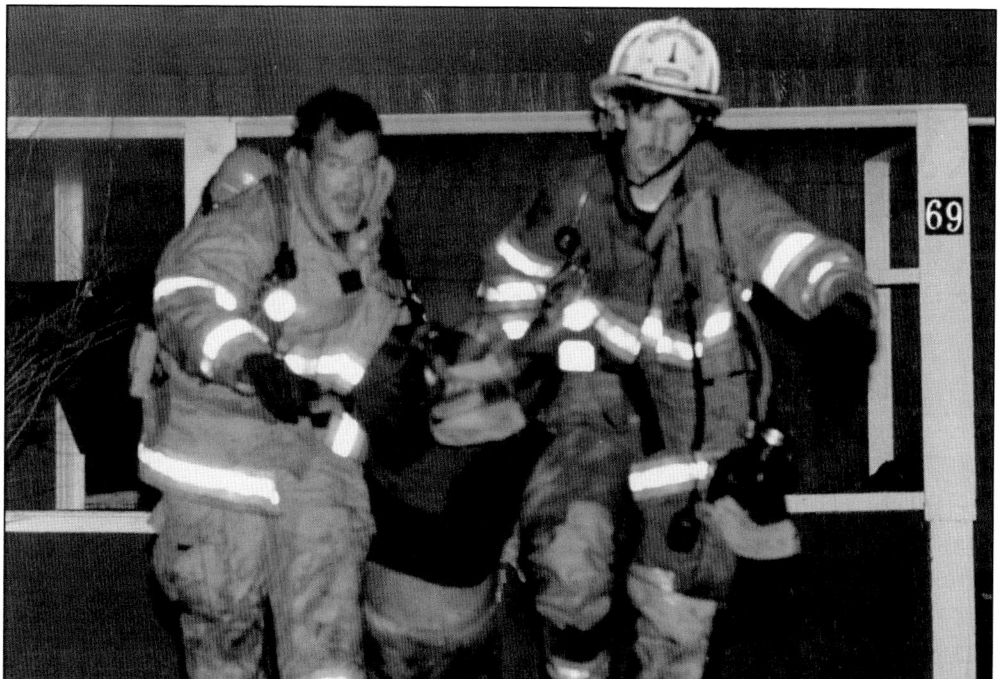

A fire that erupted in the bedroom of a first-floor apartment at 69 Prospect Street on April 7, 1995, critically injured the apartment's resident. Captured in this dramatic image are Lt. Michael Hogan (left) and Lt. Bernie Marvin (right) as they rush the victim across the front lawn to an awaiting ambulance after completing the rescue from inside the smoldering apartment.

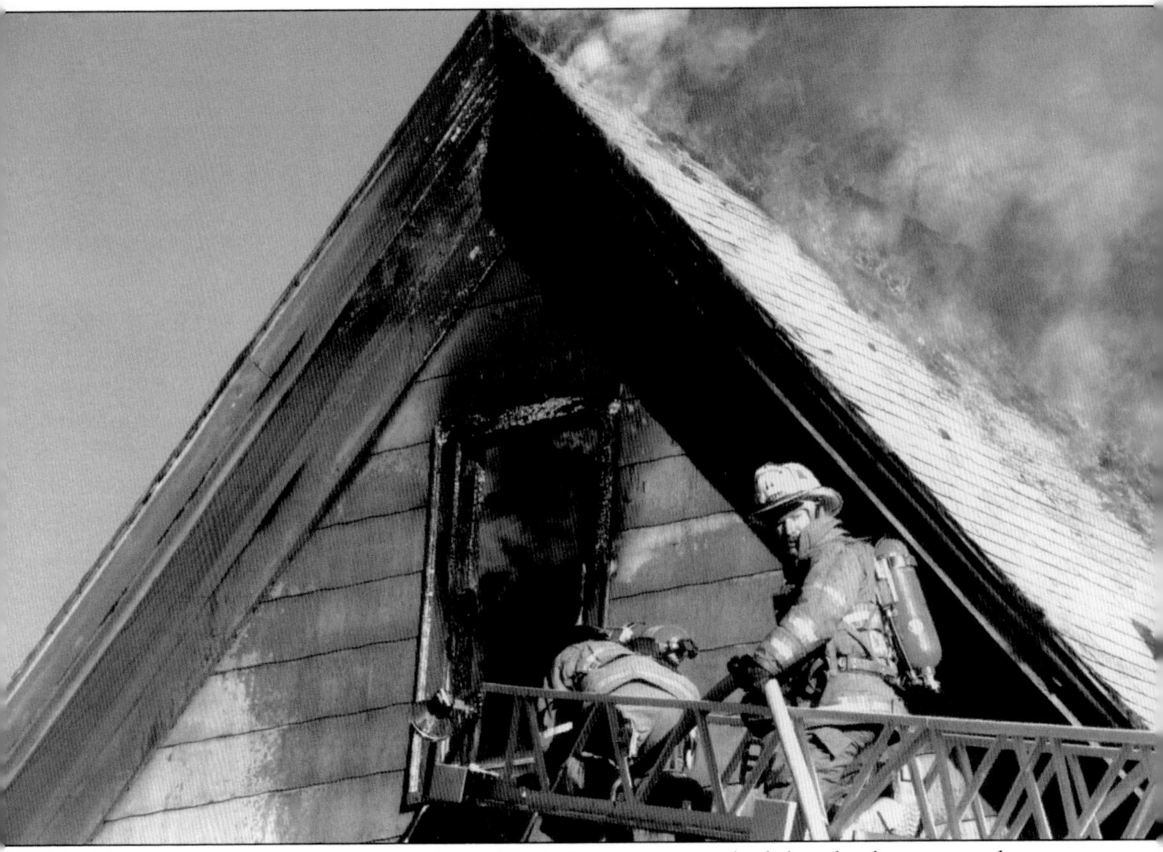

Firefighter Christopher May (left) and Capt. Bernie Marvin (right) make their way to the attic at this three-alarm blaze that destroyed the top floors of a 383 Maple Avenue home and business on December 15, 1997. Two firefighters, along with a friend of the family who rescued an occupant, were injured in the blaze. (Courtesy of the Portsmouth Herald.)

Five

For the Welfare of Others

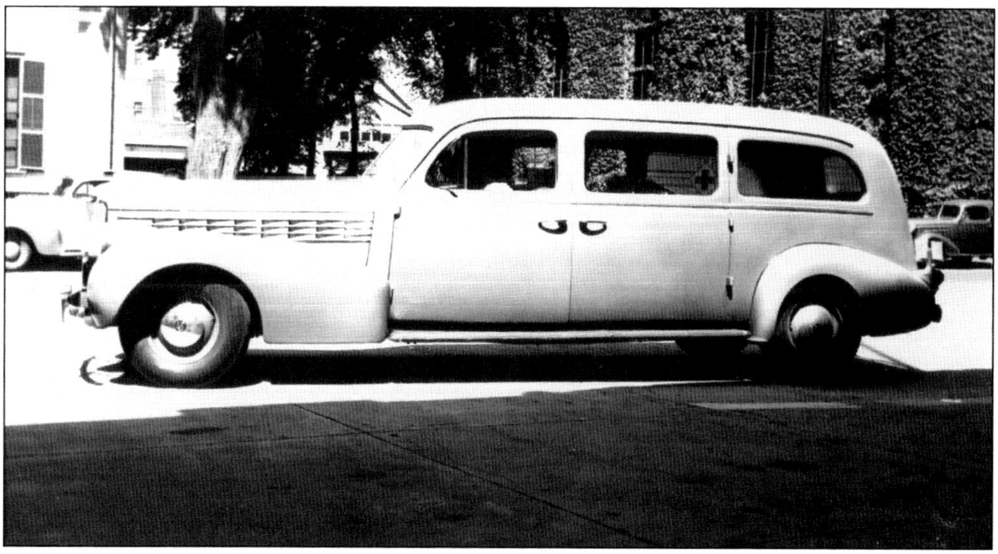

In 1944, the department purchased this ambulance and in 1945 began providing emergency medical services to the city. The annual report of the fire department for 1945 notes 367 ambulance trips, and the following year, the ambulance made 366 trips. During the first six months of 1946, the ambulance had made 223 trips. (Photograph by Francis J. Carbone, courtesy of Robert Randall.)

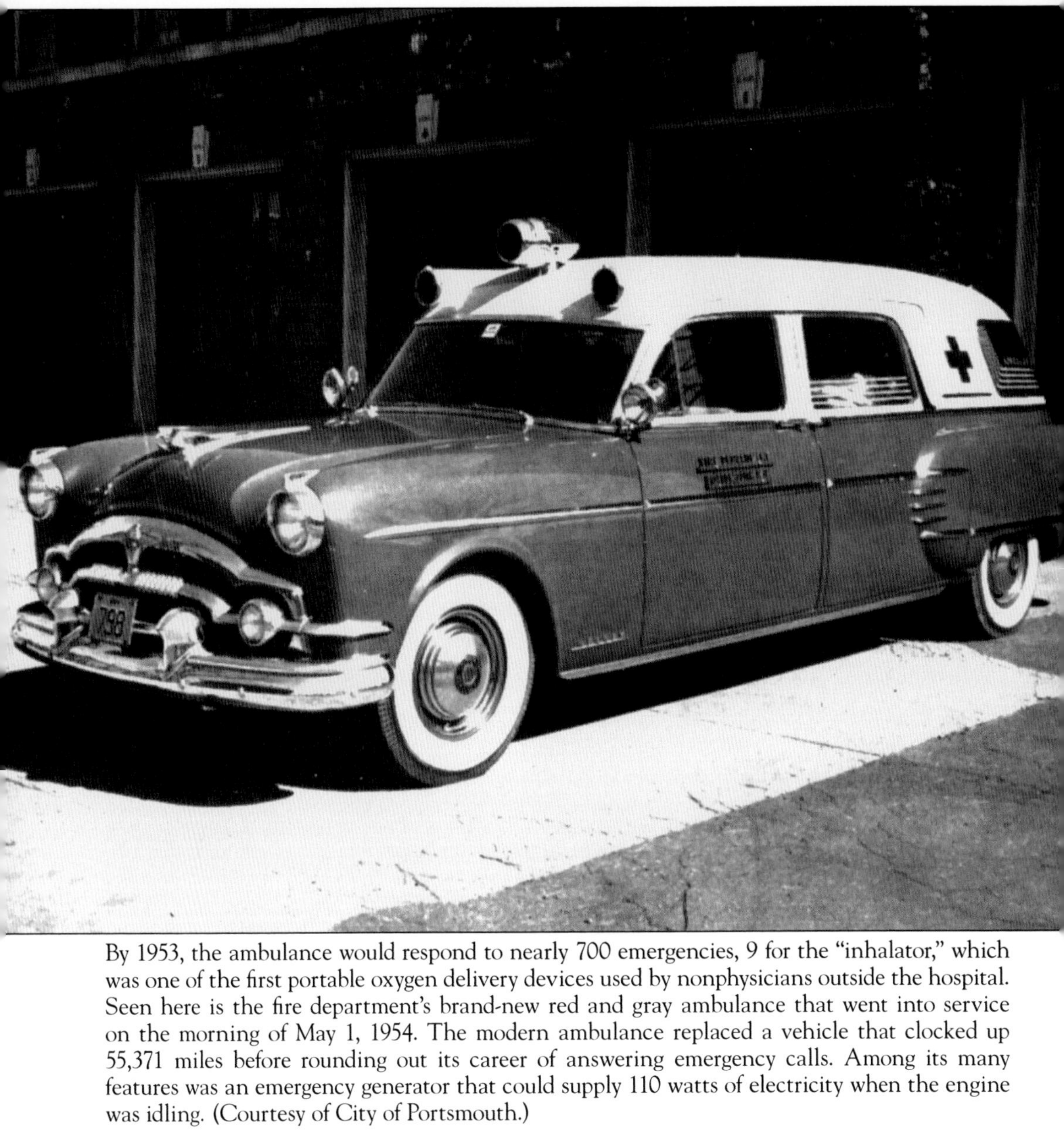

By 1953, the ambulance would respond to nearly 700 emergencies, 9 for the "inhalator," which was one of the first portable oxygen delivery devices used by nonphysicians outside the hospital. Seen here is the fire department's brand-new red and gray ambulance that went into service on the morning of May 1, 1954. The modern ambulance replaced a vehicle that clocked up 55,371 miles before rounding out its career of answering emergency calls. Among its many features was an emergency generator that could supply 110 watts of electricity when the engine was idling. (Courtesy of City of Portsmouth.)

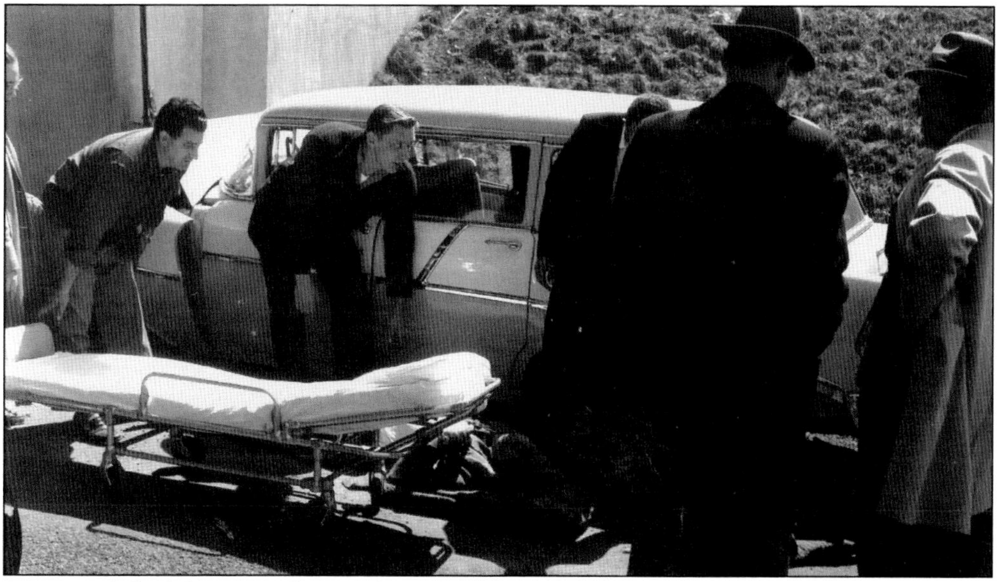

Portsmouth firefighters Nicolo Ingoglia (left) and Robert W. Orr (right) maneuver the ambulance cot through a crowd of onlookers in this photograph taken on April 15, 1958, at a car accident on the New Hampshire Turnpike. Firefighter Orr began his career with the department as a call firefighter in July 1954. He was appointed a permanent member on September 8, 1956, and retired 28 years later in 1984.

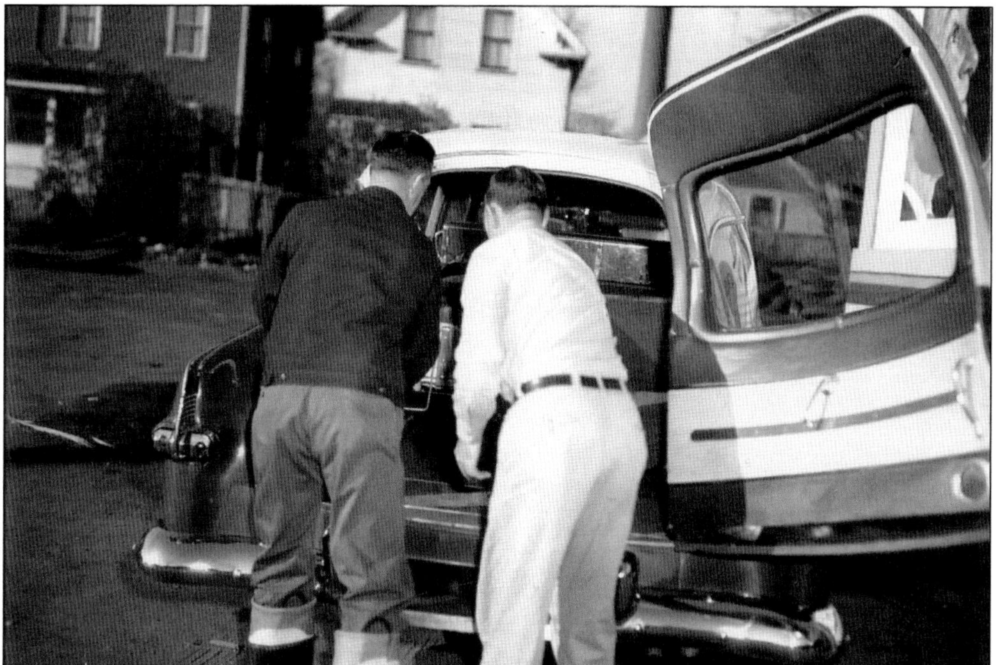

Firefighters John Downs (left) and John "Jack" Benjamin (right) are seen here unloading the gurney from the 1954 ambulance in this undated photograph. Firefighter Downs was appointed as a permanent member in 1954 and firefighter Benjamin in 1953. Both would obtain the rank of lieutenant during the late 1960s.

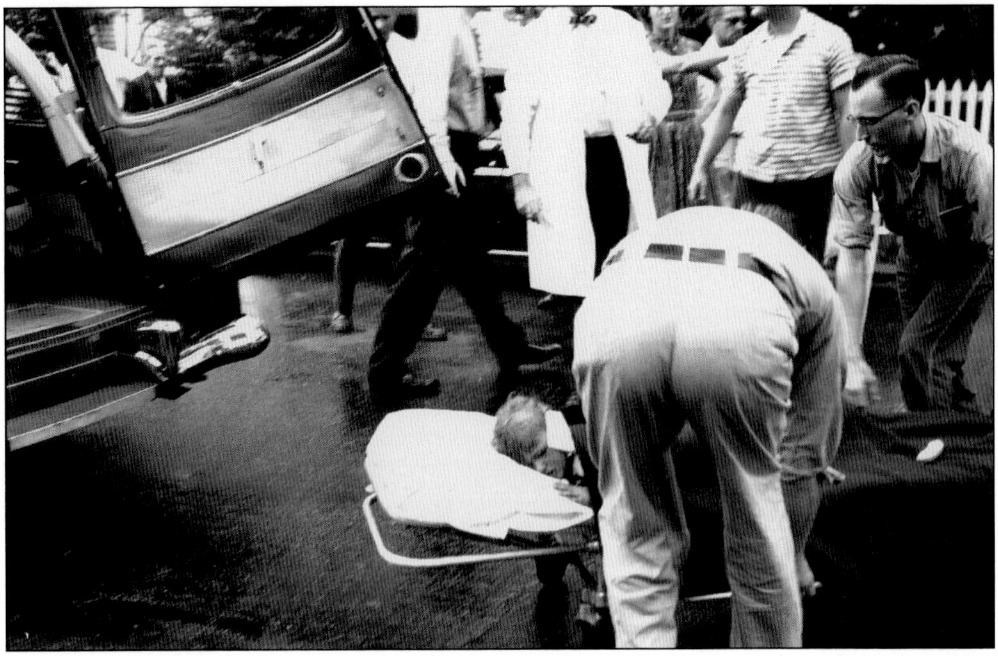

A little girl of three or four years old is being loaded into the Portsmouth ambulance in this image from September 1, 1959. According to reports, she was struck by a vehicle on Islington Street near the Central School around 1:30 p.m. and transported to Portsmouth Hospital with ear injuries and for "fright." Firefighter Frank Gillespie can be seen at the far right.

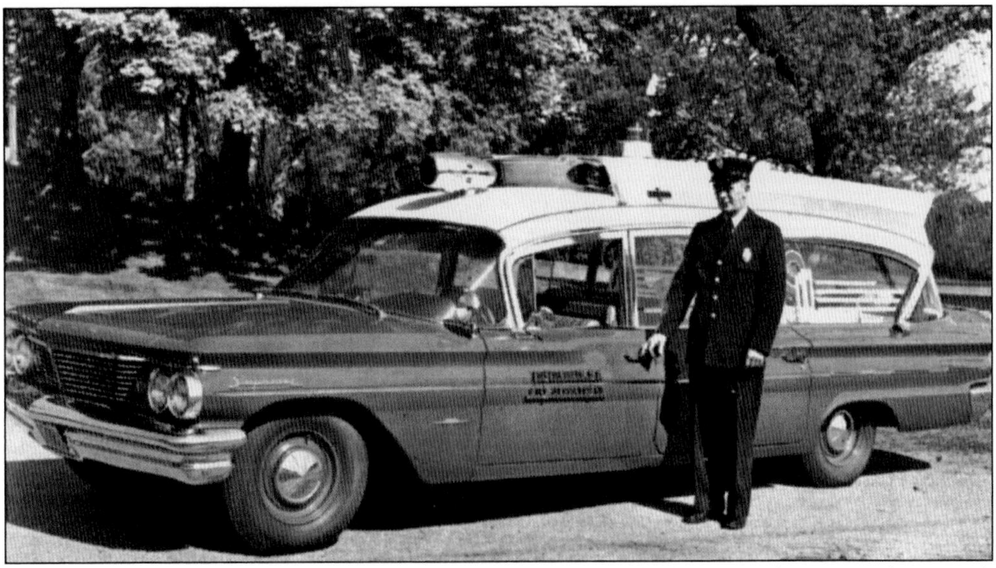

Firefighter Donald Lane stands next to the department's ambulance in this 1963 photograph from the City of Portsmouth's annual report. In that year, the ambulance would respond to 727 emergencies. Firefighter Lane first joined the department as a call firefighter for Combination No. 3 in 1935. After an absence from the department for three years starting in 1942, he was reappointed as a call man and then as a permanent fireman. Firefighter Lane would rise through the ranks, and on October 19, 1973, he became the city's fire chief. Chief Lane retired from the department in February 1976.

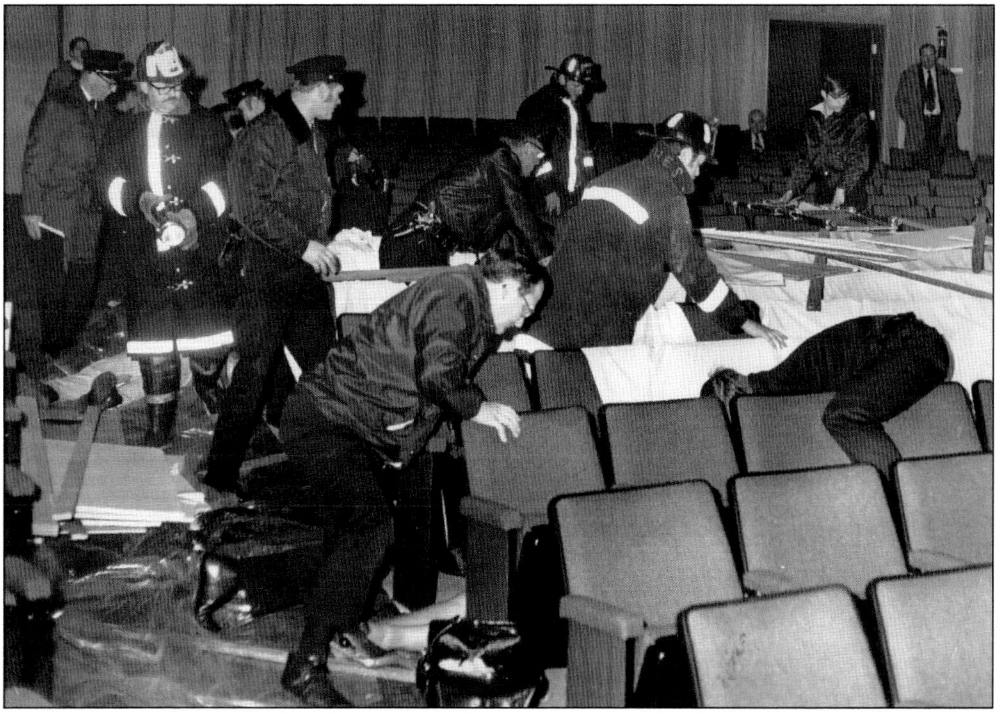

This view from inside the Jerry Lewis Cinema shows fire and emergency personnel assessing and treating the victims of a staged mass casualty accident held around 1973. Emergency personnel in the foreground of the photograph are, from left to right, Fire Chief Ernest Weeks, Fire Capt. Gene Hersey, Portsmouth police officer Joe St. Pierre, department chaplain Jack Ryans, and firefighter Robin Hughes ("HES" on collar).

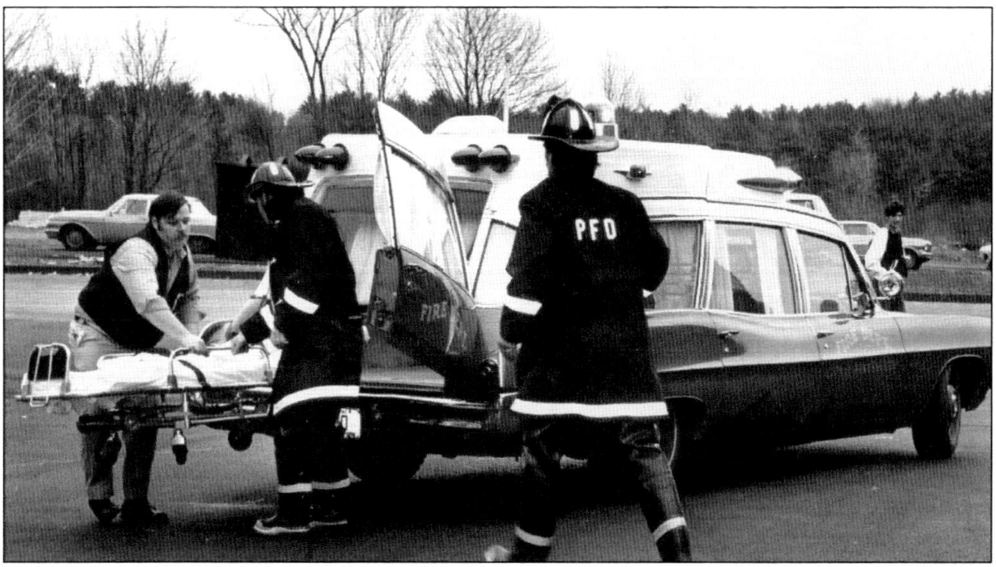

Firefighters remove the stretcher from Portsmouth's ambulance during a mass casualty drill held at the Jerry Lewis Cinema located along Lafayette Road near the present-day Bowl-a-Rama in this photograph taken sometime around 1973. Numerous volunteers playing the role of injured victims were transported by ambulance to Portsmouth Hospital as part of the exercise.

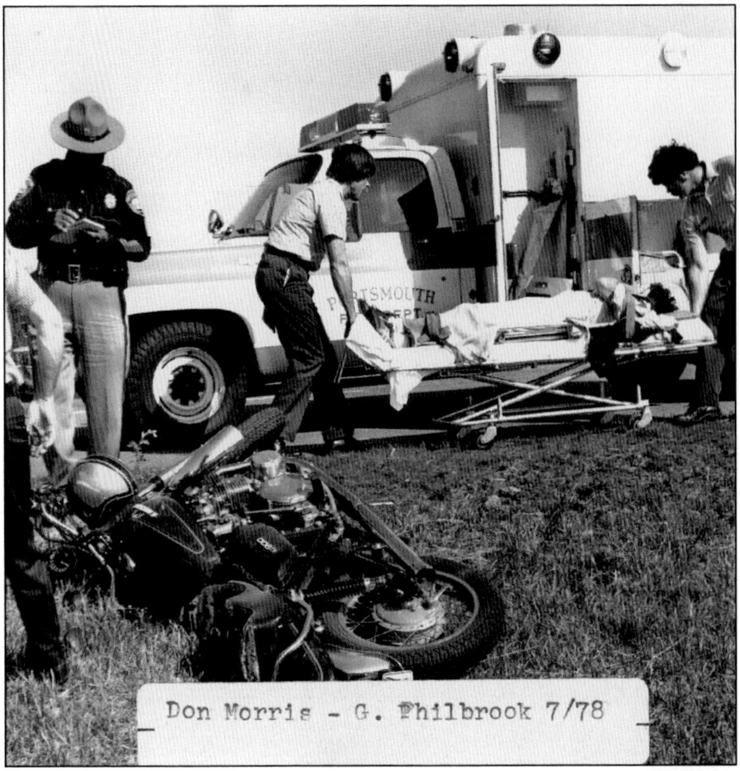

In 1976, the department began to train personnel to the EMT level. By 1978, all personnel were participating in CPR training every six months, and those that were certified as EMTs had to complete 24 hours of refresher training. In this photograph of a motorcycle accident from July 1978 are, from left to right, an unidentified New Hampshire state trooper, firefighter Donald Morris, and firefighter Gregg Philbrook.

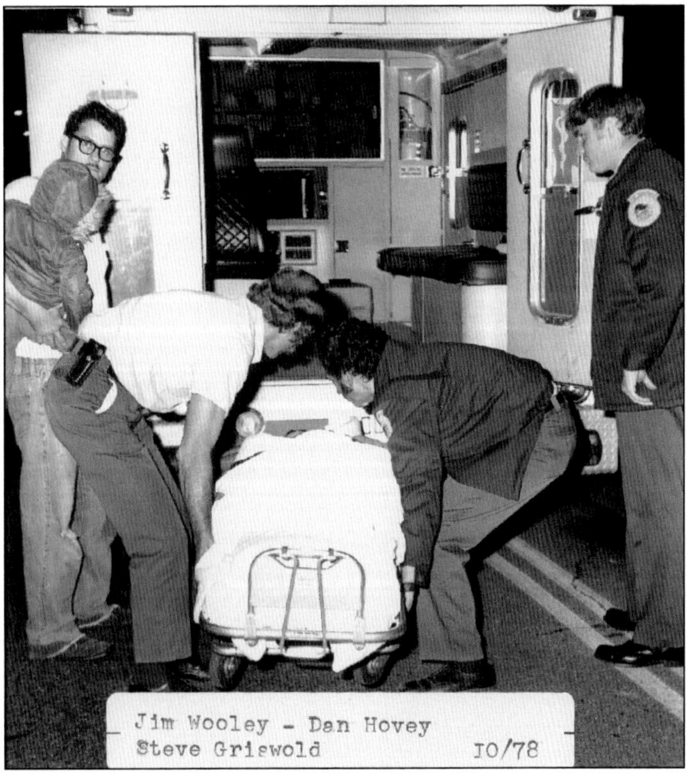

A patient of a motor vehicle crash on the Route 1 bypass is being loaded into the ambulance by Portsmouth Fire Department personnel in this image from October 1978. Pictured are, from left to right, an unidentified man involved in the accident holding his child, firefighter James Wooley, firefighter Daniel Hovey, and firefighter Steven Griswold.

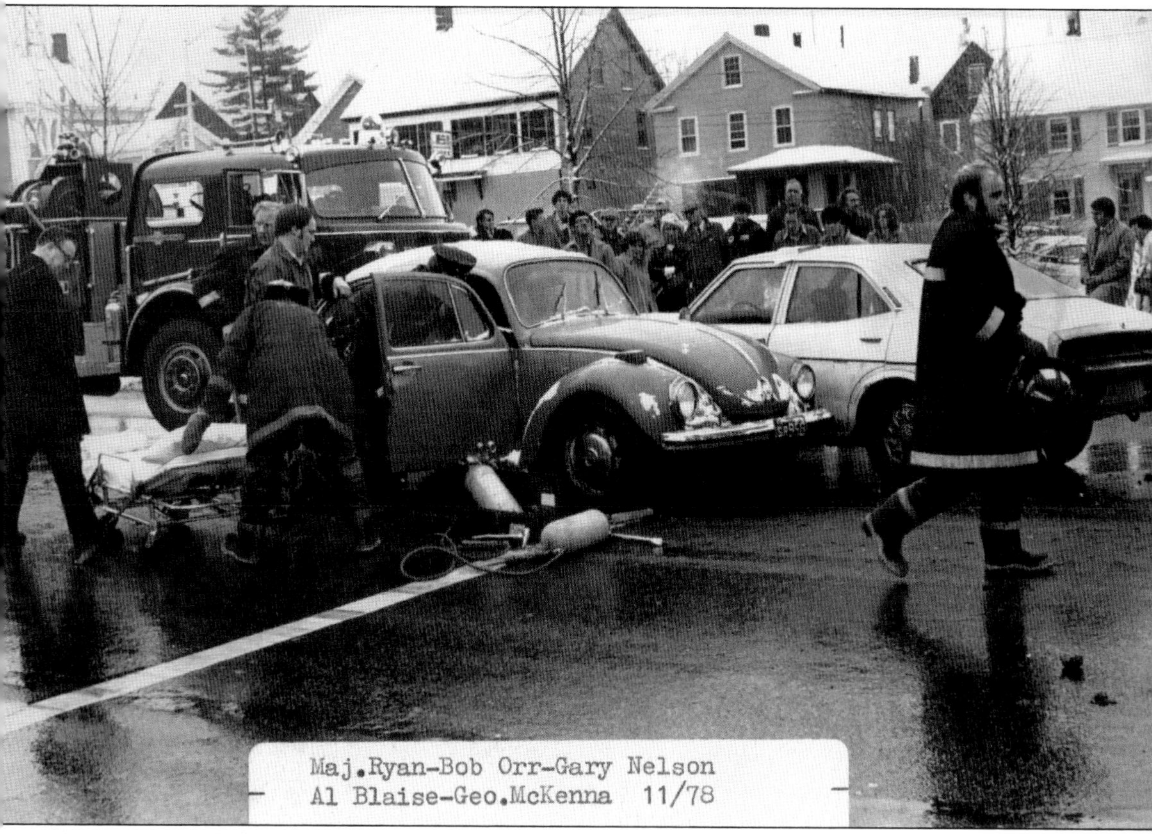

An air-powered chisel and a pry bar rest next to a Volkswagen as emergency crews prepare to remove a 25-year-old Dover woman who was trapped inside her car after a two-car accident. The November 28, 1978, crash occurred at the intersection of Maplewood Avenue and Hanover Street just before 2:00 in the afternoon. The woman was transported to Portsmouth Hospital for muscle strains. The driver of the other vehicle was uninjured and cited for running a red light. Personnel in the photograph are, from left to right, chaplain Maj. Jack Ryans, Bob Orr, Garrett Nelson, Al Blais, an unidentified police officer, and George McKenna.

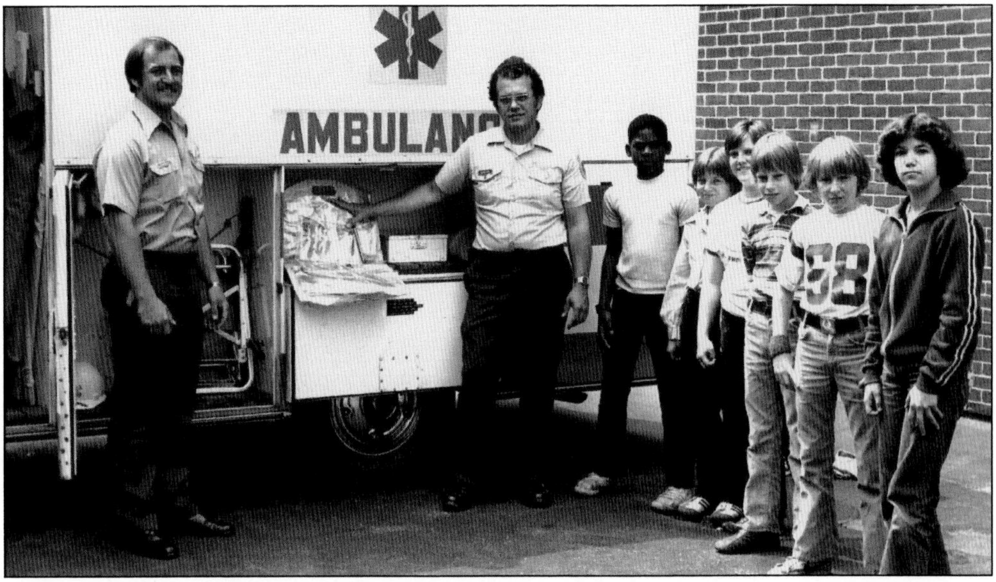

Firefighters Kevin J. Kelly (left) and Alfred (Al) G. Blais (right) introduce students from the Portsmouth Junior High School to the equipment carried on the ambulance in this photograph taken in June 1979. Firefighter Kelly would serve the city for just under five years beginning in 1978. Firefighter Blais, however, retired at the end of 2000 with over 23 years of service as a permanent member of the department.

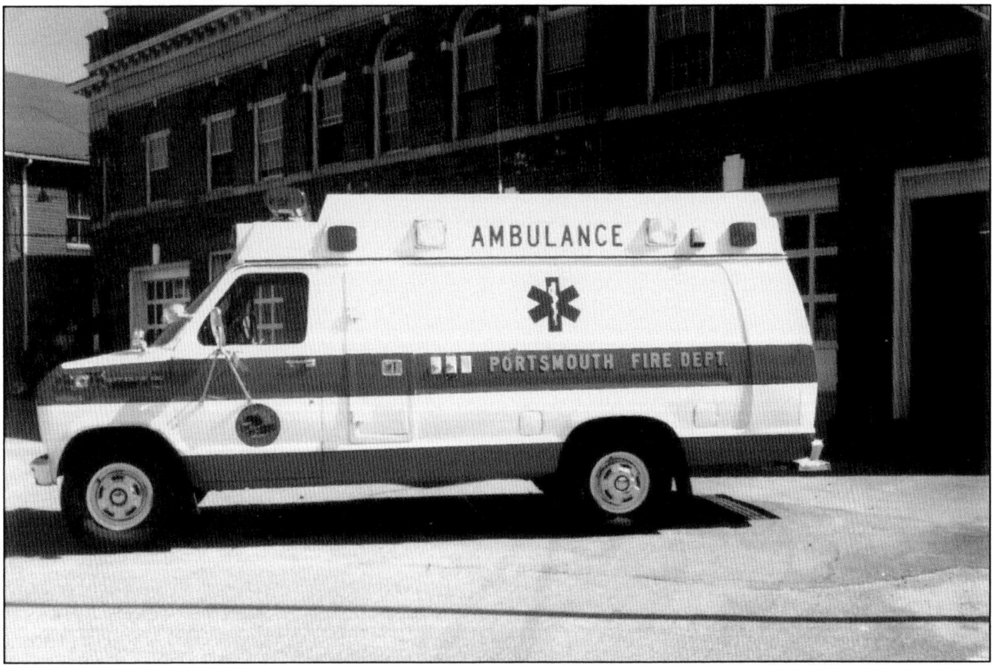

By 1987, the department boasted 20 EMTs. All remaining personnel were Red Cross Advanced First Aid qualified with subsequent training in patient assessment, oxygen therapy, and vehicle extrication. This photograph of the new 1986 Ford E-350 XL van ambulance was taken at the Central Fire Station on Court Street in April 1987. The ambulance was delivered to the city two months earlier and replaced a 1982 Super Van ambulance.

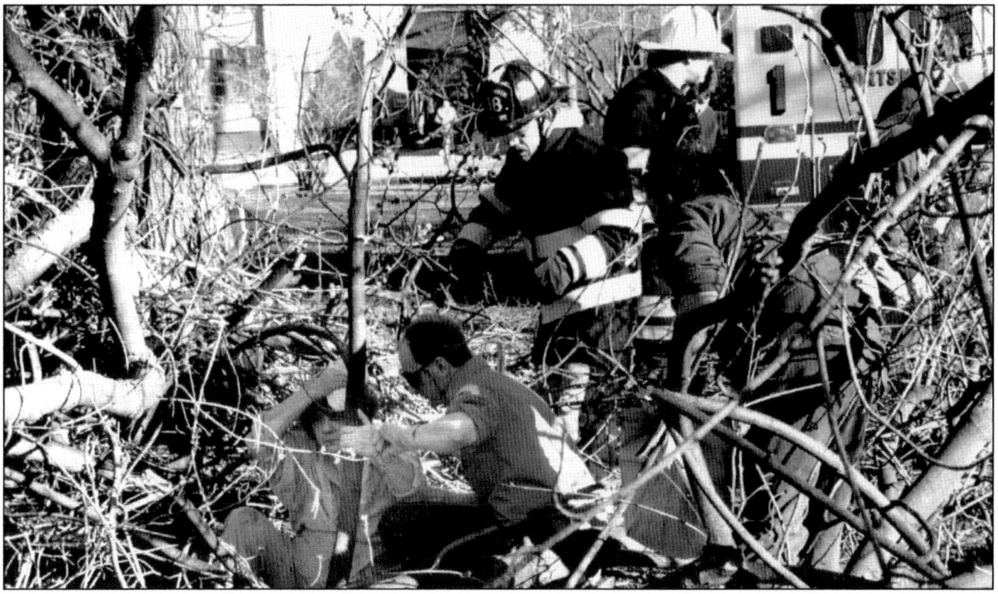

Firefighter James Heinz is seen in this March 28, 1991, image treating a woman who was injured when a tree came crashing down at Goodwin Park along Islington Street as a result of heavy winds. Also in the photograph are, from left to right, firefighter Al Blais, Capt. Steven Griswold (white helmet), firefighter Tom LaMontagne, and firefighter George "Butch" Francois. (Courtesy of the Portsmouth Herald.)

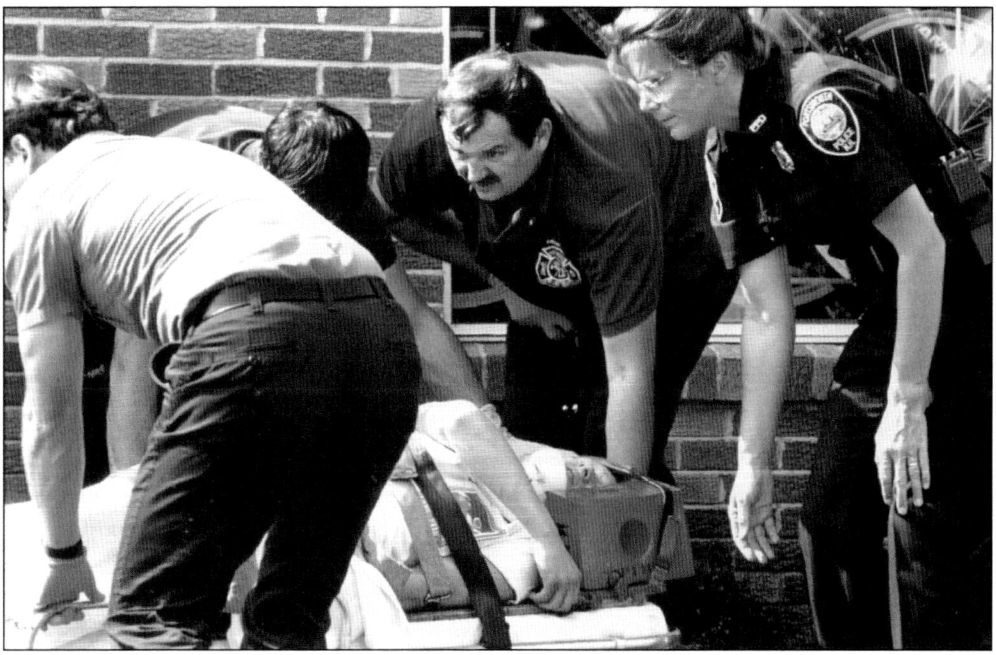

Portsmouth firefighters move a victim of an Islington Street accident to the ambulance in this August 1992 photograph. Firefighter and EMT Dennis Pease is in the center of the image managing the head of the stretcher. Unidentified members of the fire department and Portsmouth police officer Karen Johnson (right) assist in the movement of the patient to an awaiting ambulance.

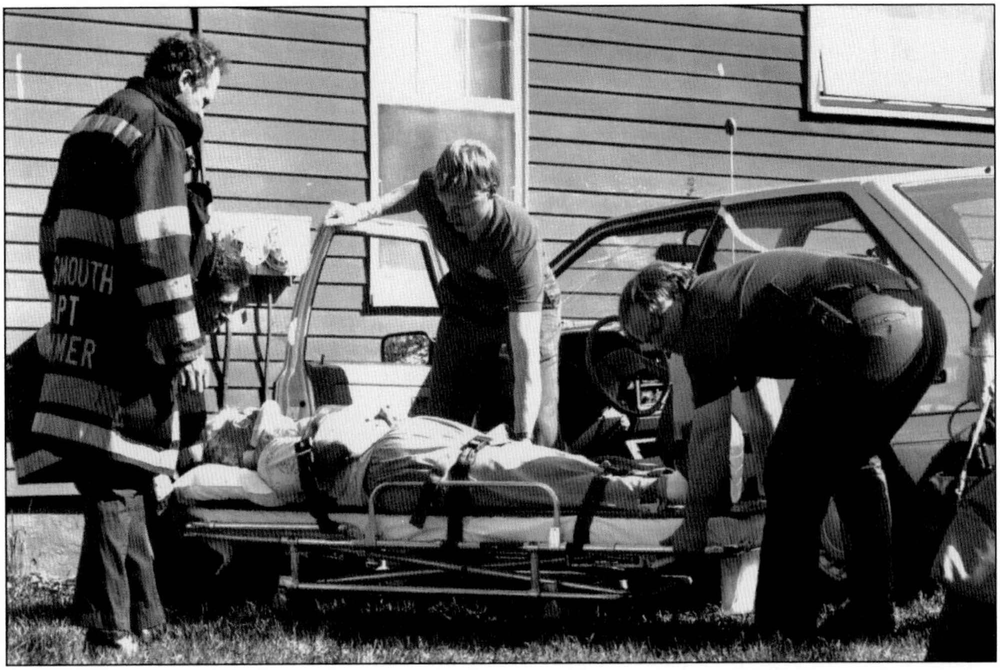

Portsmouth fire and emergency medical crews prepare to move the elderly driver of a car that crashed into a house to the ambulance. Seen here in this photograph taken around 1994 are, from left to right, Capt. Ricky Plummer, Charles Cormier, Steve Smith, and Richard Murphy. Plummer would later be promoted to chief of the department, leading the organization from 1996 to 2001.

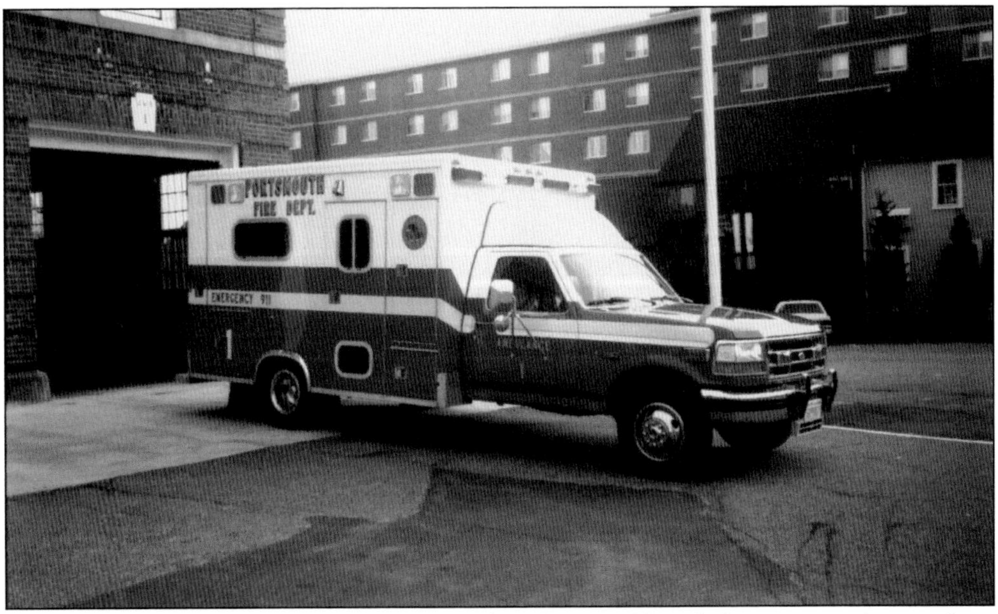

This 1999 image of Ambulanc 1, a 1995 Ford F-350 Type III ambulance, was taken in front of Station 1 on Court Street. That same year four members of the department enrolled in a one-year intensive paramedic training program. The following year, paramedic-level ambulance service began in the city.

Six

SPECTACULAR BLAZES

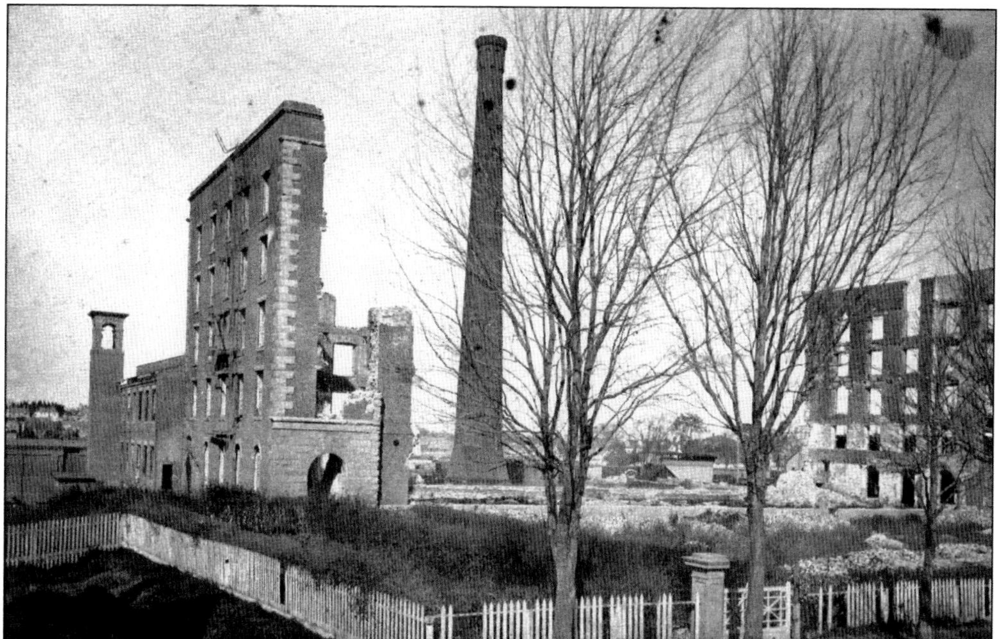

On Saturday, December 4, 1880, the Kearsarge Mill, formerly the Portsmouth Steam Factory, was destroyed by what was reported as the worst public calamity since the great fire of 1813. The fire ignited in the southeast corner of the fifth story, and before the blaze was under control the buildings were almost entirely destroyed. The Pearl Street Church also suffered damage, largely due to water. A number of other buildings some distance from the conflagration, including the Baptist church on Middle Street, were set aflame by the flying embers, although most were extinguished immediately.

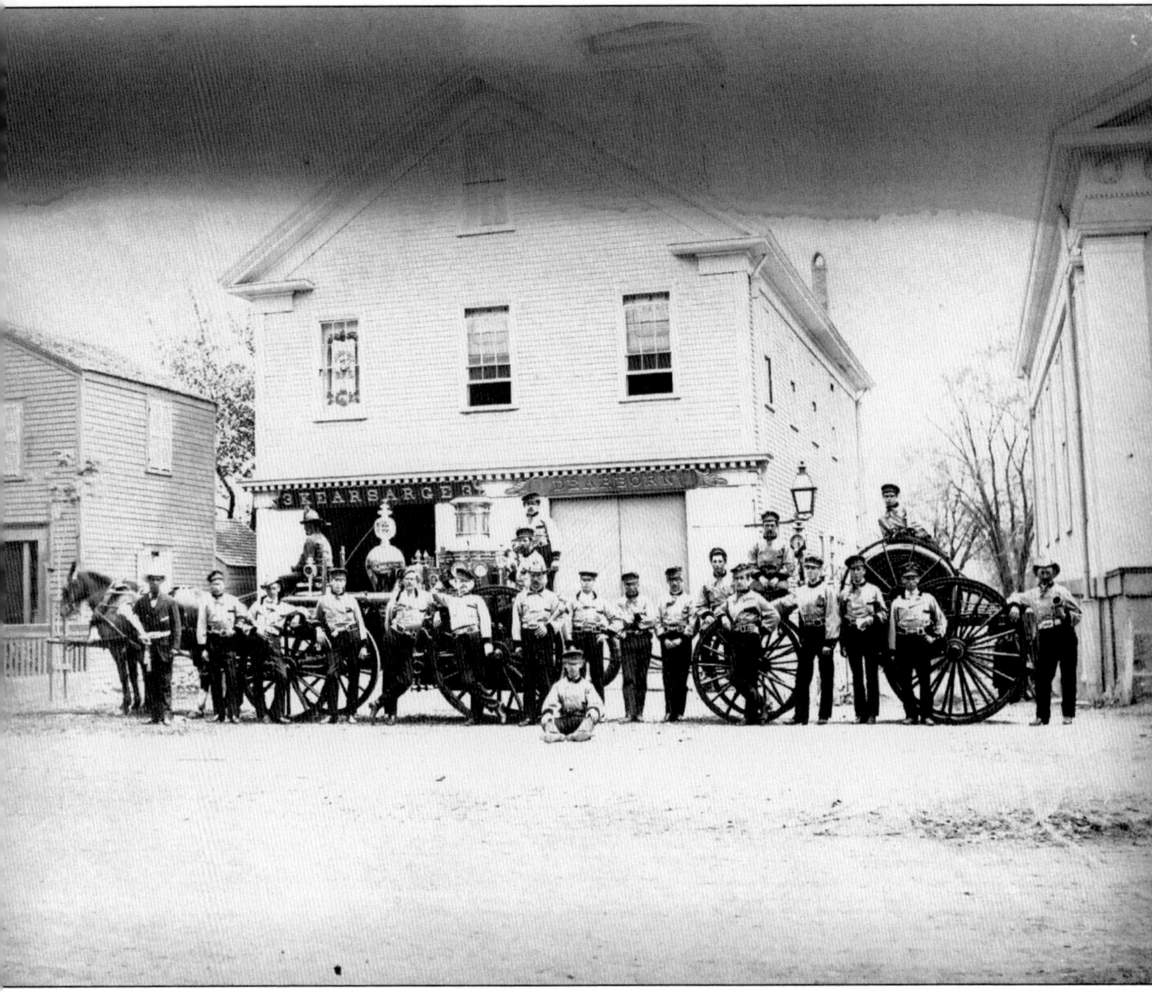

On Sunday, November 10, 1872, at 1:30 a.m. the captain of the night watch received a telegram from the mayor of Boston, asking for assistance from Portsmouth to respond to what would be later known as the "Great Boston Fire of 1872." Portsmouth mayor Thomas Marvin, as soon as notified, answered "Will come immediately with a steam fire engine, meet us on arrival." At 2:05 a.m., the alarm was sounded, and 25 men, along with Kearsarge No. 3 and a detail of 9 men from Dearborn SFE, boarded the Boston and Maine Railroad and headed for Boston. Seen here in this photograph taken during the early 1870s are members of the Kearsarge SFE Company along with their steamer (left) and hose wagon (right) in front of the Central Engine House.

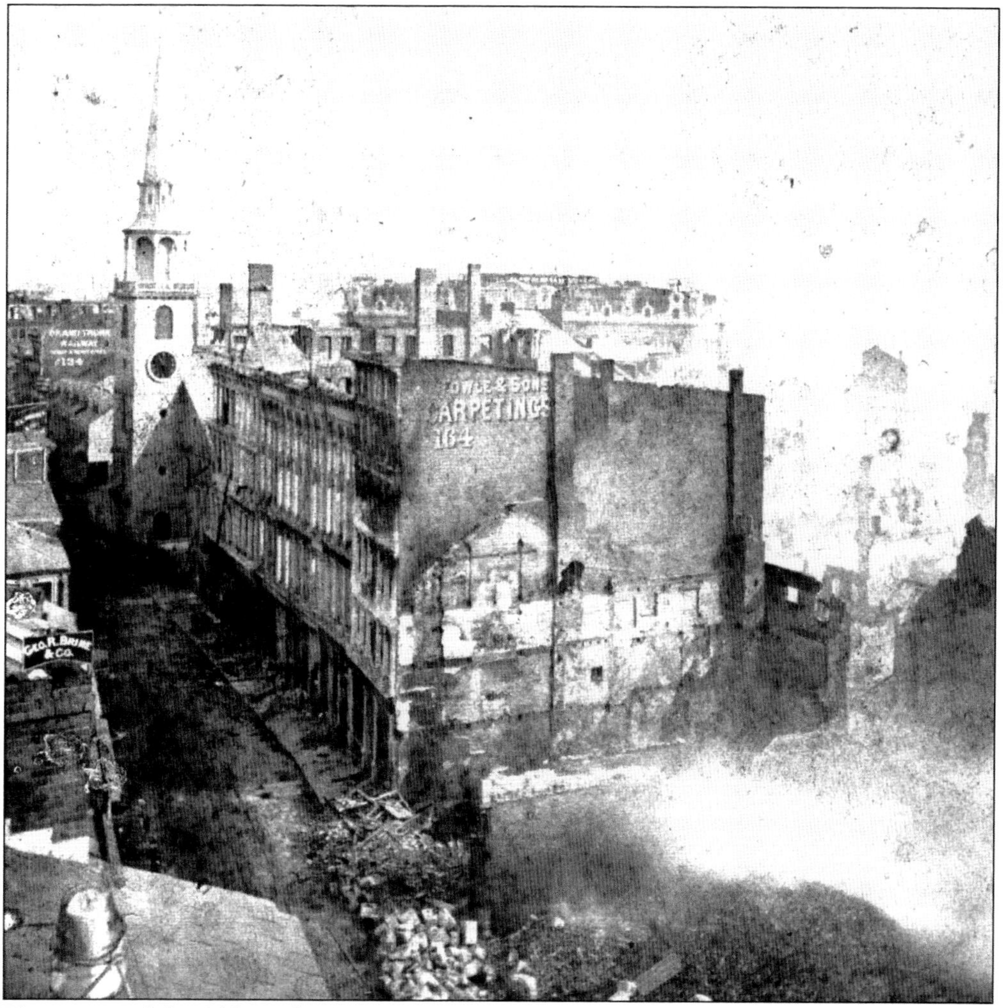

This image of the Old South Church still standing among the ruins along Washington Street was taken by Charles Taber and Company of New Bedford, Massachusetts. As recollected by Harold Murdock in *Letters Written by a Gentleman in Boston to His Friend in Paris*, "a single steamer was sending up a column of black smoke that silhouetted against the glow beyond. They were driving her hard, as she quivered and roared she seemed pitted alone against a world of flame. The Old South had evidently been through a drenching, and as I stepped across to ask the engineer man where he hailed from, I read 'Kearsarge, Portsmouth, N.H.', on the boiler plate."

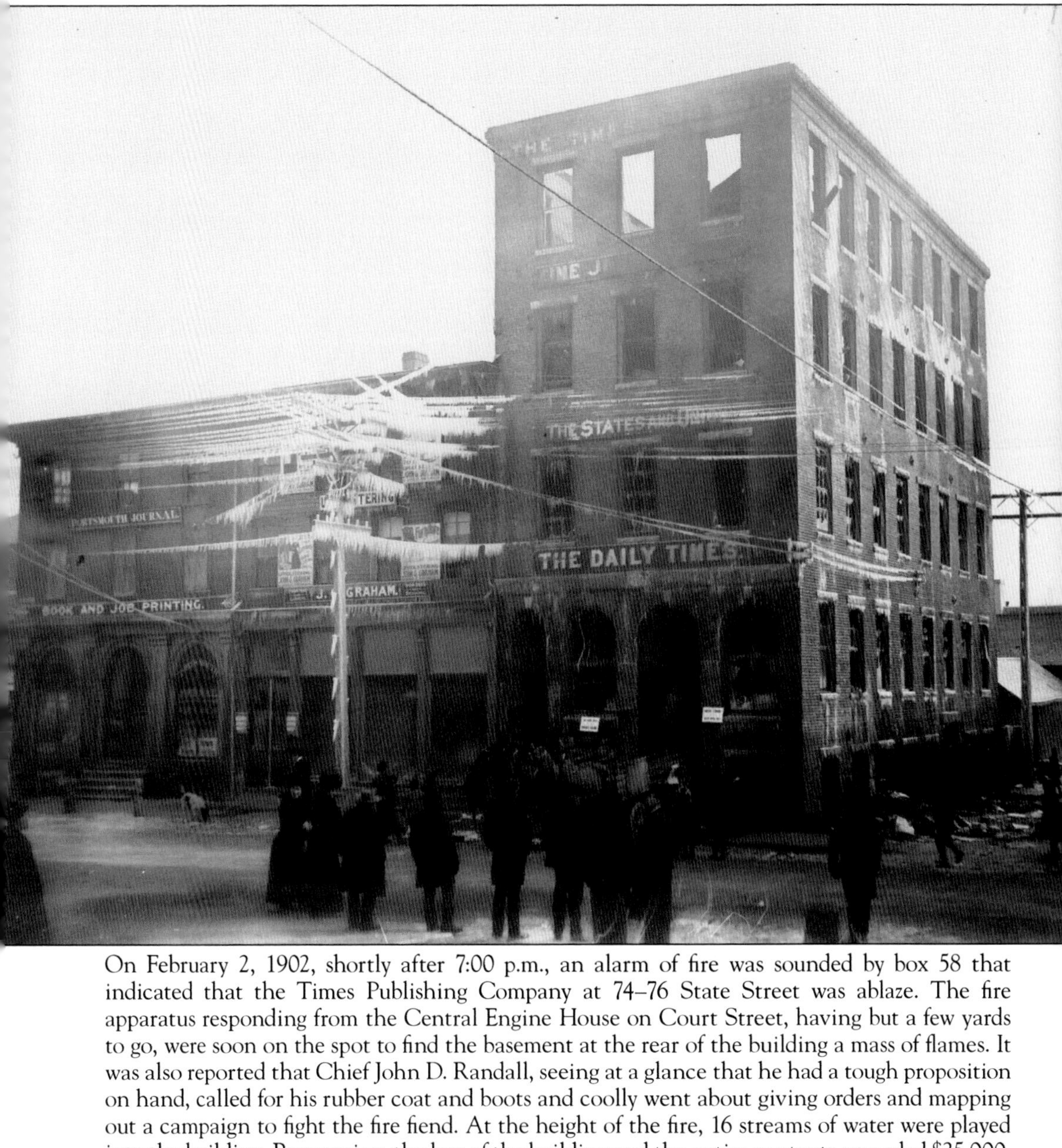

On February 2, 1902, shortly after 7:00 p.m., an alarm of fire was sounded by box 58 that indicated that the Times Publishing Company at 74–76 State Street was ablaze. The fire apparatus responding from the Central Engine House on Court Street, having but a few yards to go, were soon on the spot to find the basement at the rear of the building a mass of flames. It was also reported that Chief John D. Randall, seeing at a glance that he had a tough proposition on hand, called for his rubber coat and boots and coolly went about giving orders and mapping out a campaign to fight the fire fiend. At the height of the fire, 16 streams of water were played into the building. By morning, the loss of the building and the entire contents exceeded $25,000. (Courtesy of Strawbery Banke.)

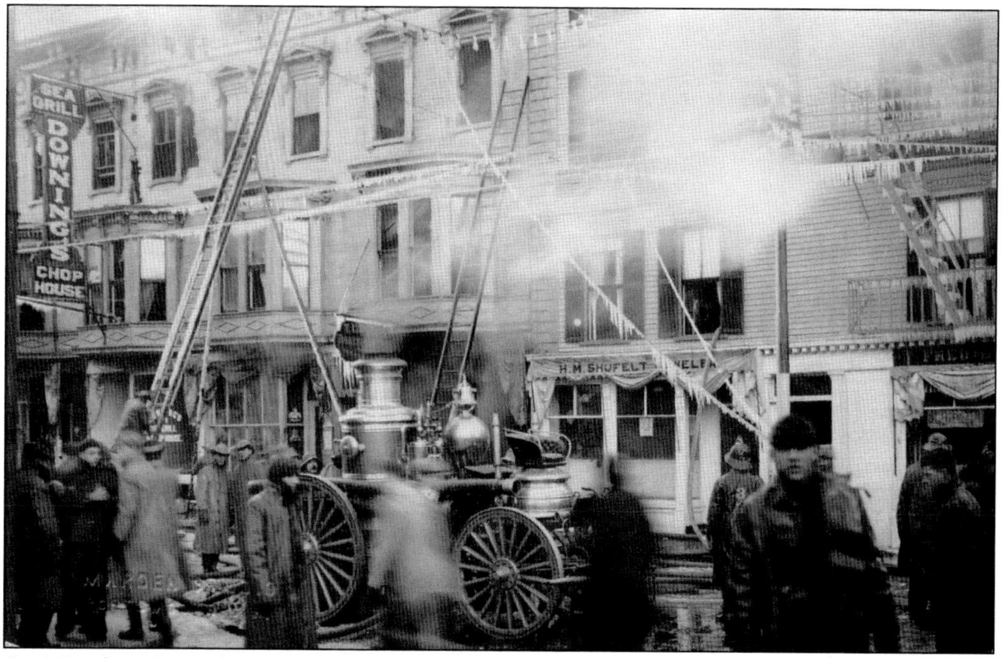

On Monday, November 23, 1914, two of the worst fires in years broke out within a few minutes of each other. The first was a fire in the machine shop and roundhouse at the railroad yard at the corner of Cabot and Islington Streets. Both were completely destroyed, and the sparks blown by the wind threatened that entire section of the city. During this fire, another blaze was discovered in the wooden block occupied by the Peabody Music Company, Downing's restaurant, Dondero's fruit store, and Dolan's cigar store in the area of 111 Congress Street. All buildings were heavily damaged, and at one time, the Langdon Hotel seemed to be the next victim. These images are of the Congress Street fire.

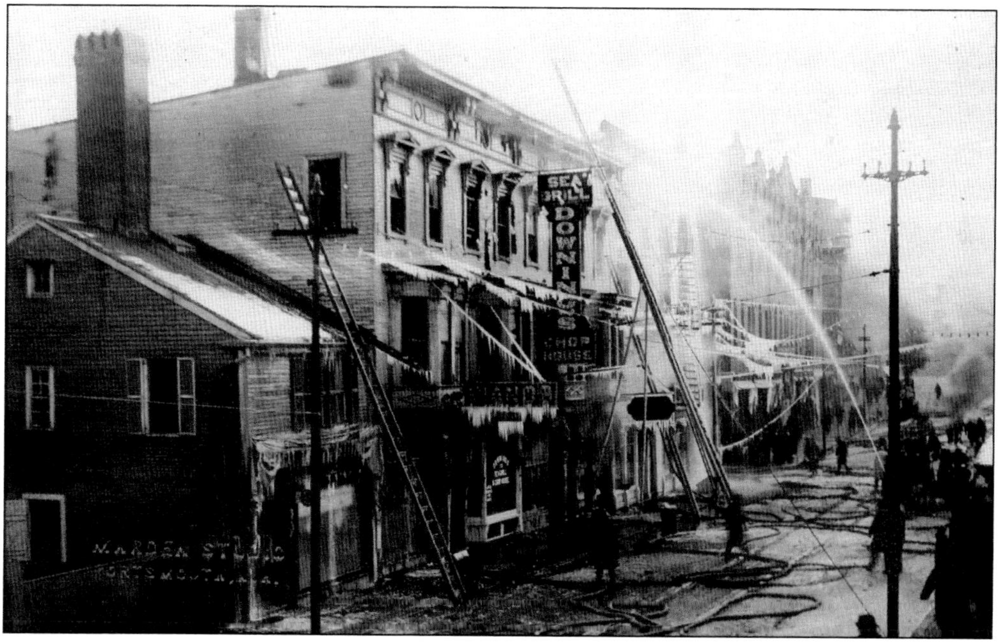

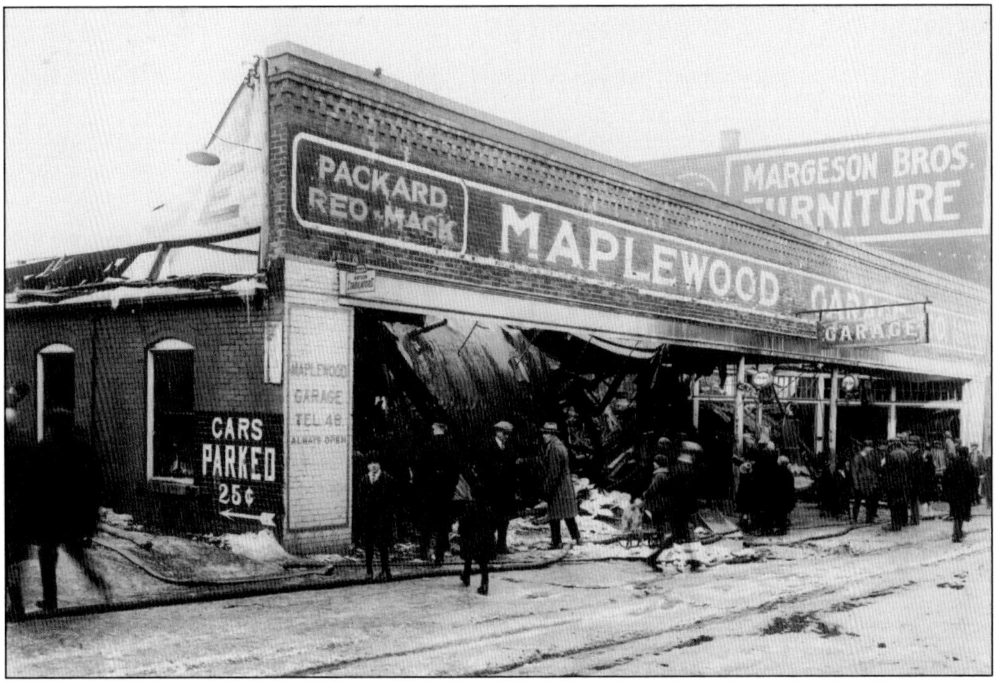

A general alarm fire on Tuesday night of March 2, 1926, at the Maplewood Garage on Vaughn Street near Hanover claimed the life of Harold Haddock, destroyed over 100 automobiles, and threatened the business section of the city. Haddock was asleep in his car close to where the fire started and was burned to death. The building was a mass of flames and spread to surrounding structures. Showered with sparks, the Olympia Theater was in the direct path of the flames, but firefighters stationed on the roof quickly extinguished all embers. It was reported that the fire lit up the sky for miles and could be seen as far away as Wells, Maine. The blaze marked the fourth fire to occur on the site of the garage over a 50-year period.

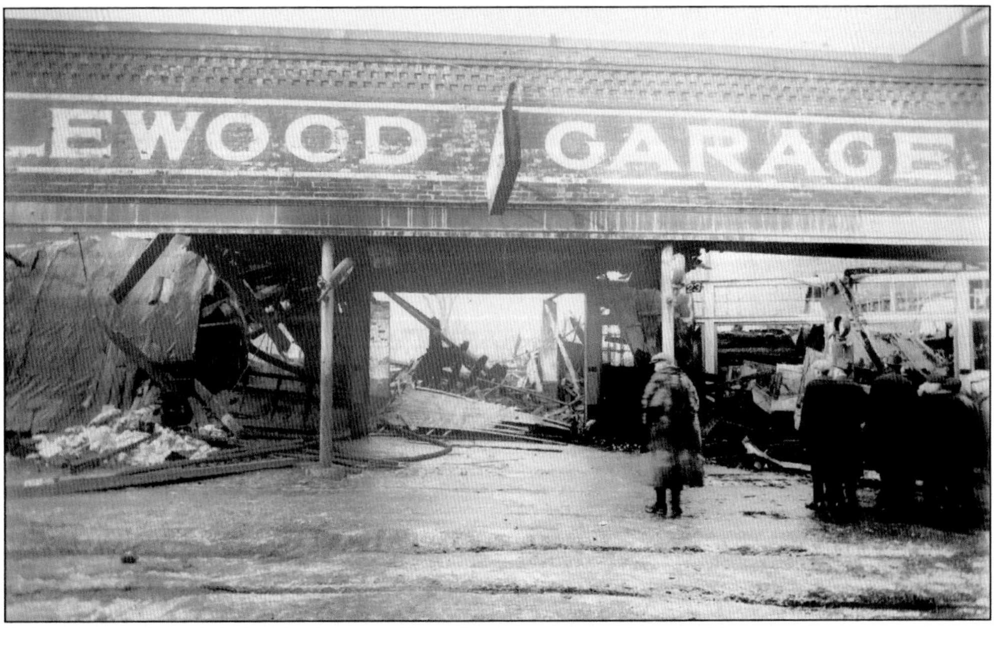

A fire that was discovered around 2:00 p.m. in the McIntosh building on Congress Street on May 5, 1931, caused one of the most disastrous fires in years. The fire started in a paper chute in the rear of the building and quickly spread through the structure. At the height of the fire, blazing embers covered the north end as the entire section of the city from the business district through the north end was threatened with destruction. It was reported that over 20 roof fires were in progress at once, and every hydrant in the business section was in operation. The buildings used by Consolidated Coal Company were also gutted. Assistance was called in from the Portsmouth Naval Shipyard, Dover, Hampton, York, Kittery, and Eliot fire departments.

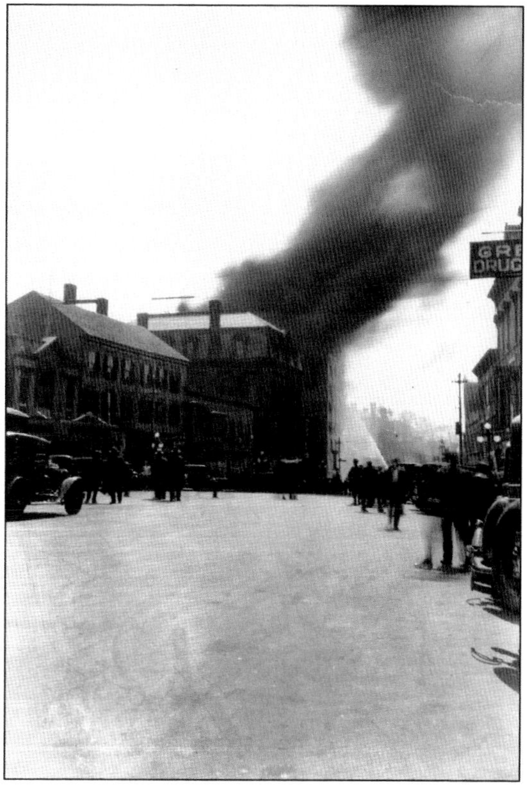

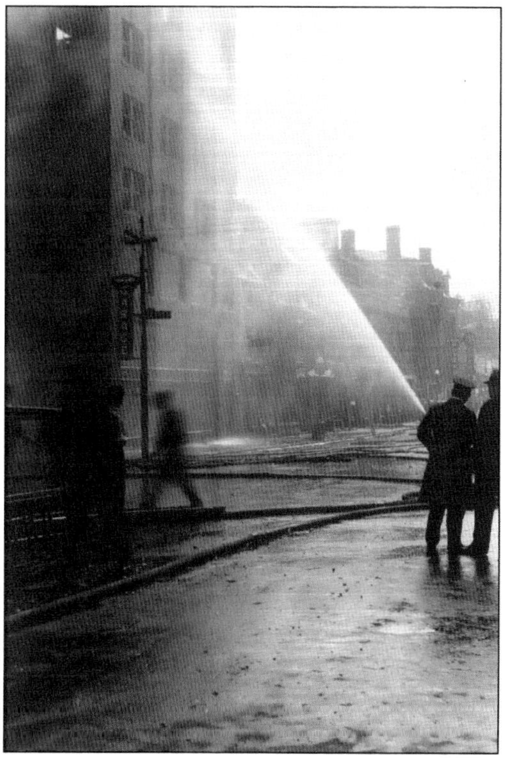

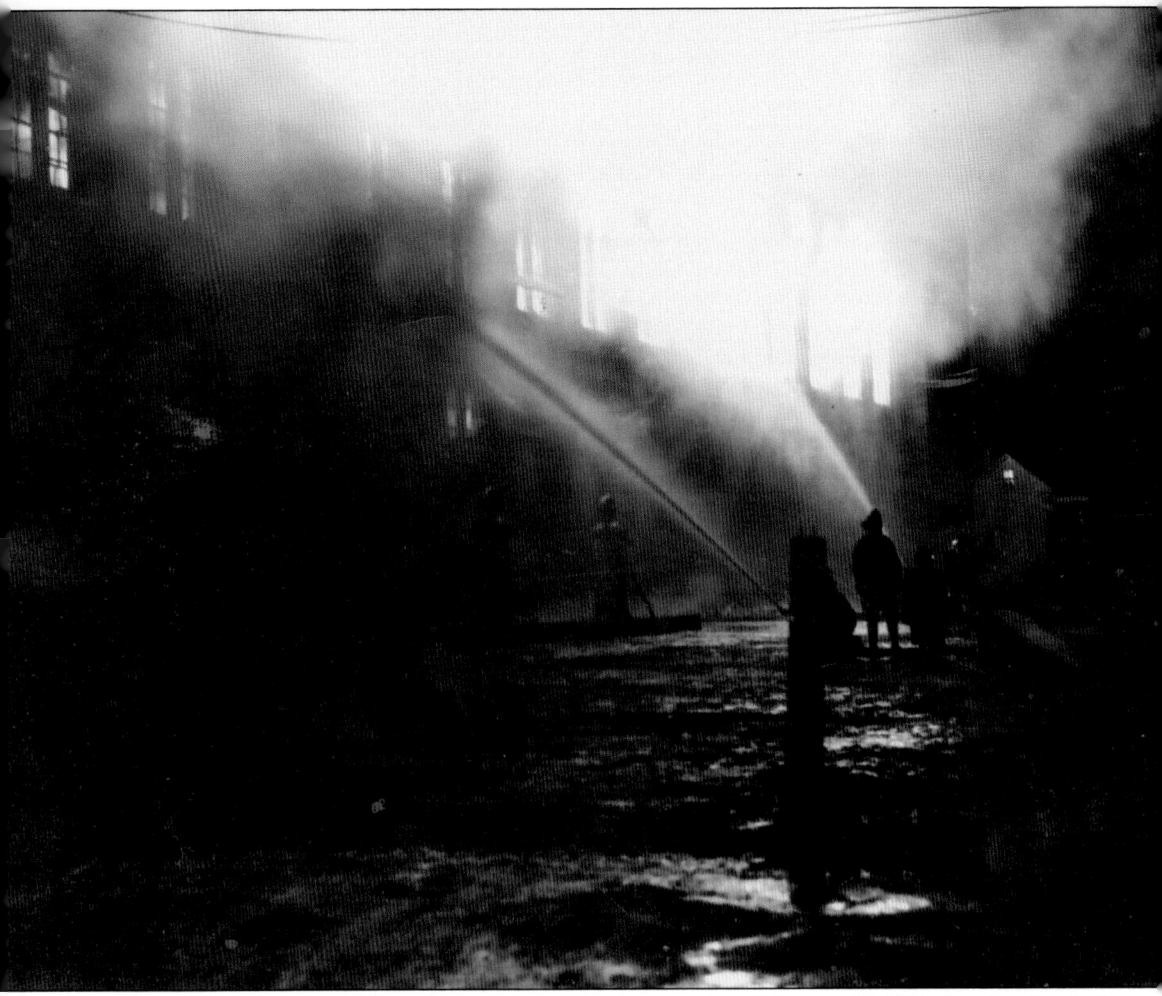

Chief George T. Cogan blamed spontaneous combustion for a February 20, 1939, fire that leveled three large buildings and threatened an entire section of the city from Hanover Street to the Piscataqua River. The blaze started within the Diamond Match Company Plant, spread through the Sussman Cleaning Plant, and fanned by a southwest wind, burned into Brooks Motor Sales. Workmen in Brooks worked frantically to move about 100 cars out of the building as flames curled toward the building. Reports indicate that in the wake of the destructive fire that caused an estimated $175,000 damage, only a few fire-baked brick walls stood, wire pipes were twisted into a maze of steel, and 15 automobiles were left charred and mangled in the ashes of the garage. Additionally, thousands of dollars worth of dry lumber and planing machinery was lost besides the valuable clothes-pressing and cleaning equipment. Over 20,000 spectators gathered from the seacoast of New Hampshire, southern Maine, and northern Massachusetts to watch the flames and smoke as it billowed hundreds of feet in the air. Motorists driving along Route 1 were also attracted to the red glow.

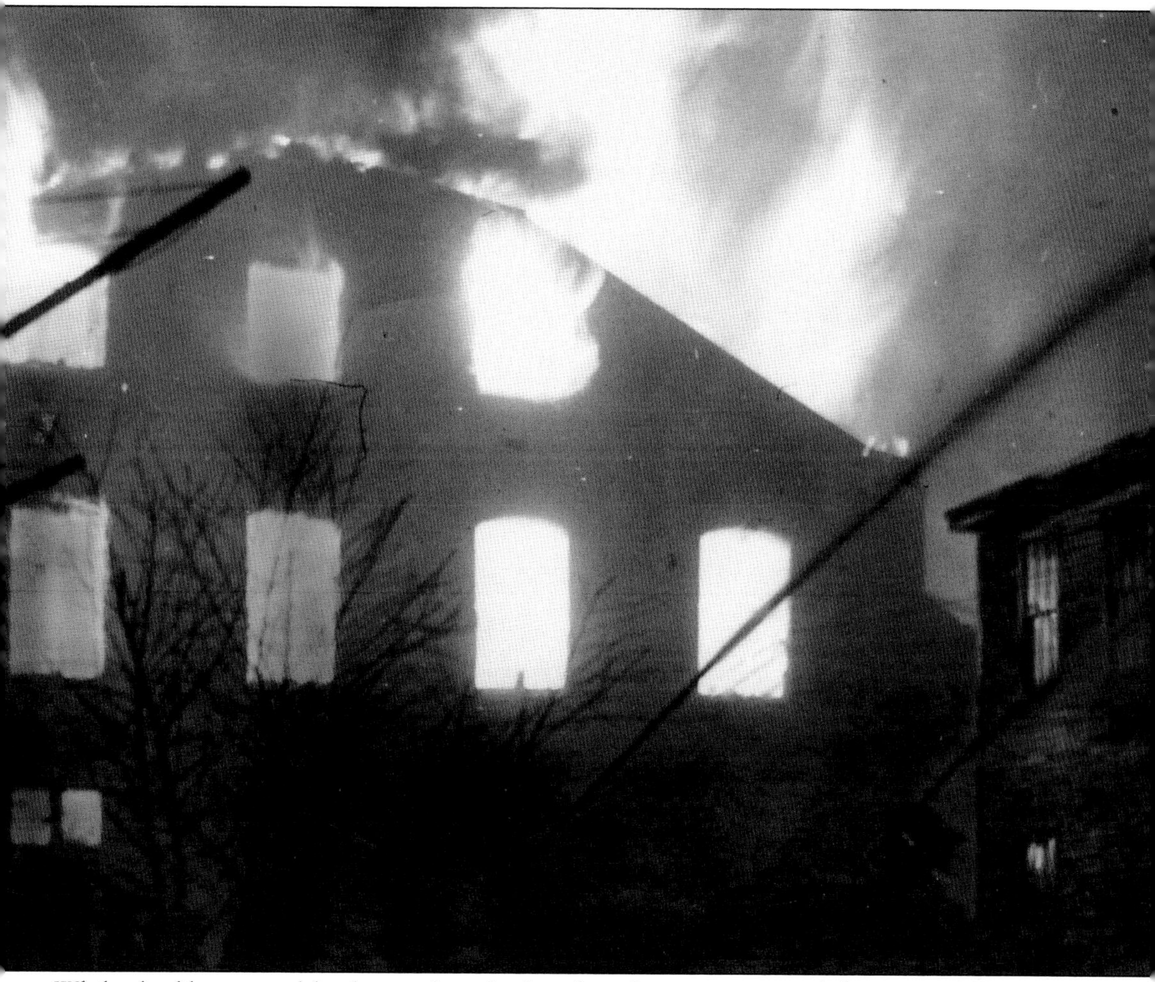

While the blaze roared loud enough to be heard in distant sections of the city and caused a reflection in the sky that could be seen for more than 25 miles, firemen from the city, the Portsmouth Naval Shipyard, New Castle, Kittery, Kittery Point, North Hampton, and Eliot demonstrated what was reported as heroic efforts to stop a blaze from spreading to the residential sections during the evening of February 20, 1939. News of the conflagration that ripped through the Diamond Match Company Plant, Sussman Cleaning Plant, and Brooks Motor Sales flashed around Portsmouth and brought people rushing from the moving picture theaters fearing for their safety and their homes. The origin of the fire in the three-story brick storehouse building of the Diamond Match Company was believed to have started around 8:00 p.m., and a few minutes later, the flames were reported as they burst through the roof. All fire danger was under control before midnight, although piles of lumber were still smoldering the following day.

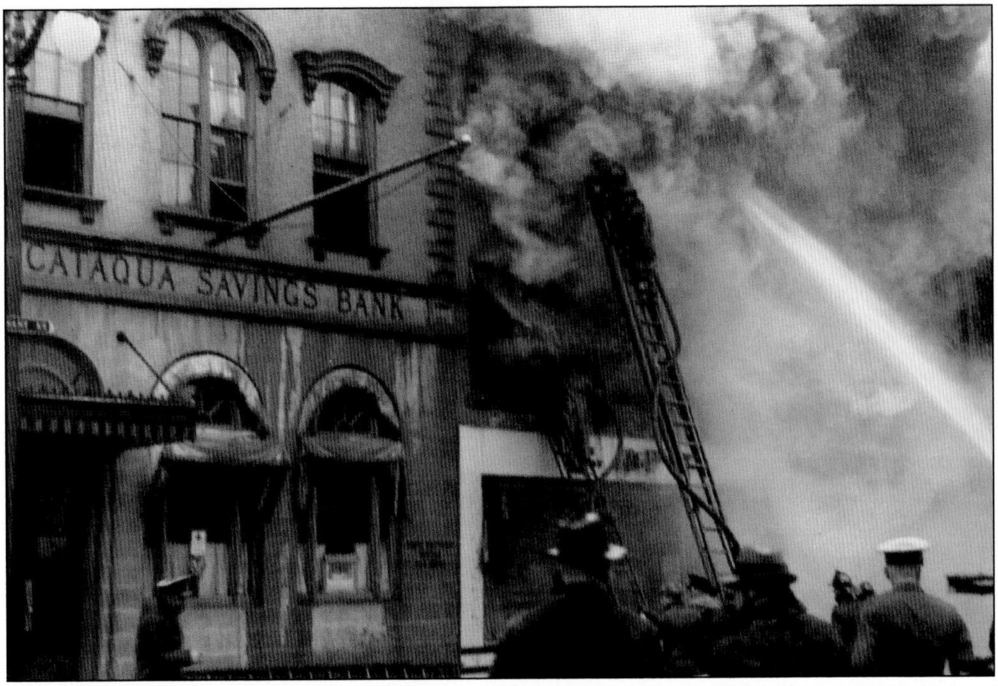

Portsmouth firefighters battle this spectacular general-alarm fire at the Exchange Block located at 21 Pleasant Street on Sunday, February 2, 1941. Five occupants were rescued over ladders that were thrown both at the front and rear of the buildings. Hampered by the lack of an aerial ladder truck, firemen had to wait for the fire to burn through the slate roof before any significant progress could be made in extinguishing the flames. (Courtesy of Robert Randall.)

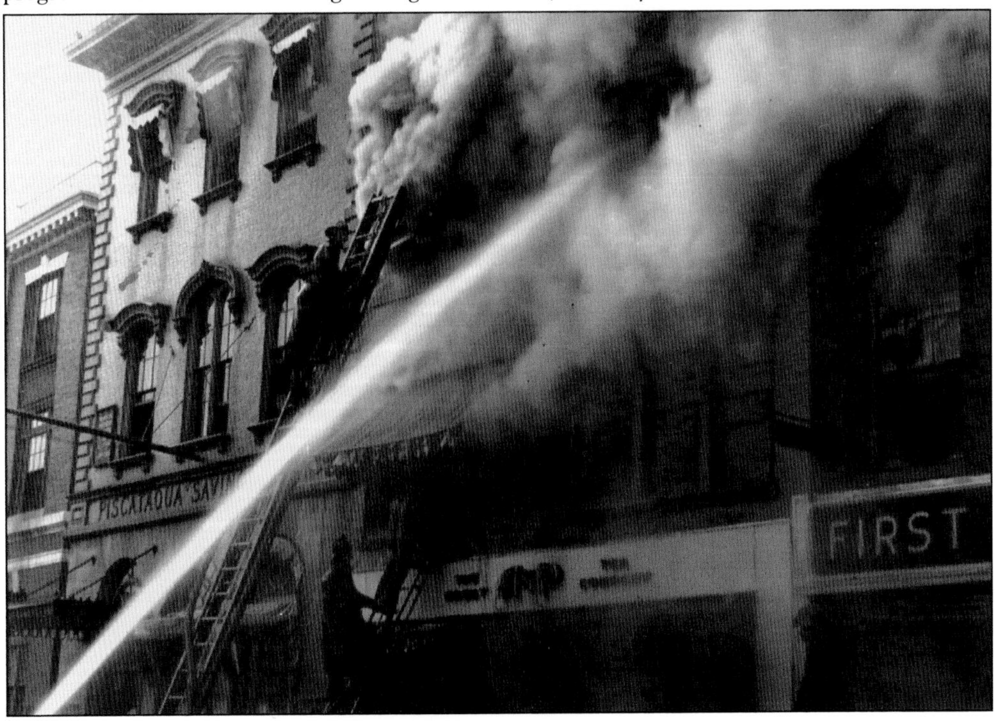

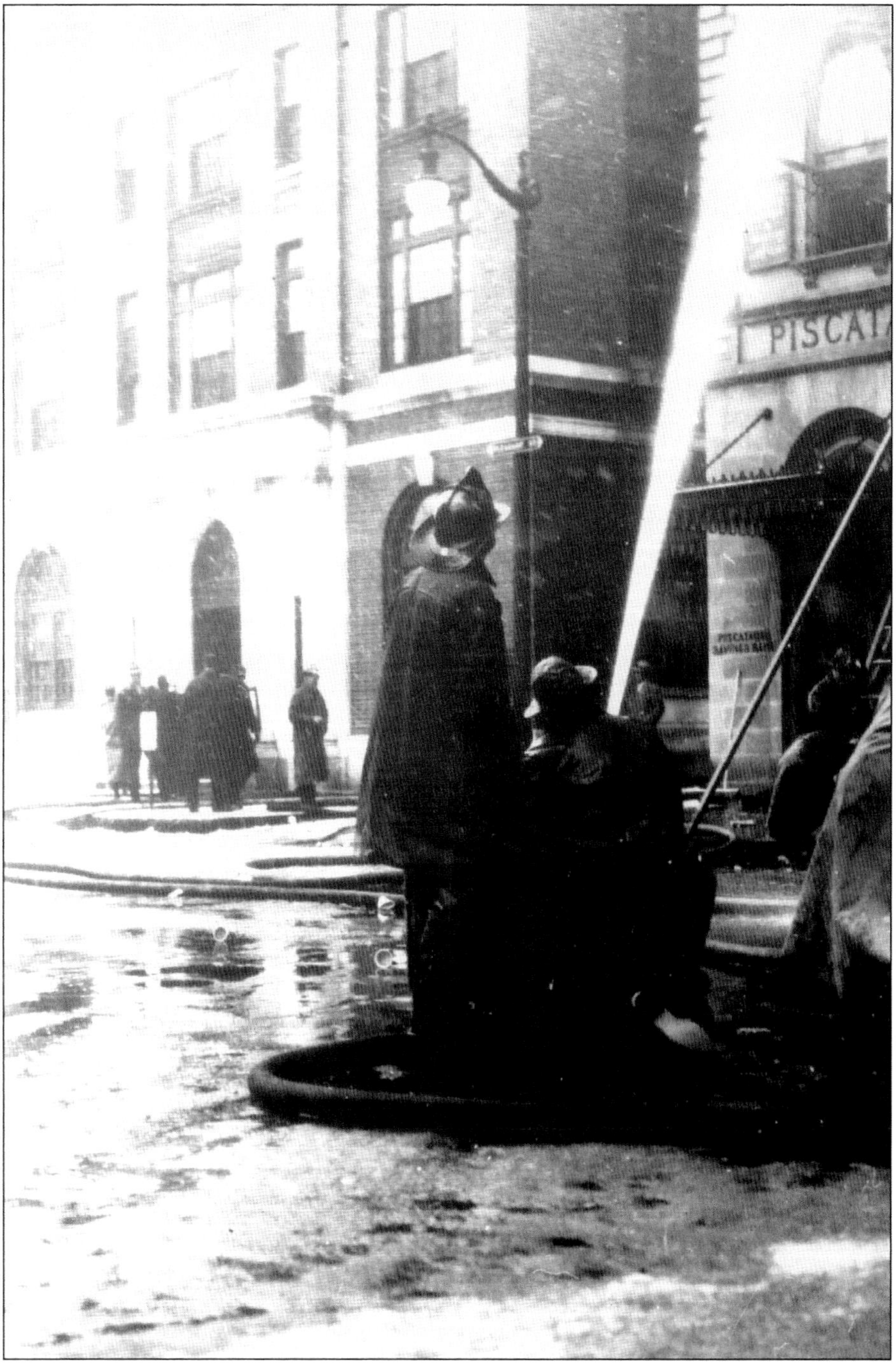

In this photograph taken by Anne Randall Culberson, William I. Randall (left) of Engine Company No. 3 oversees the application of a master stream by an unidentified firefighter (right) at the Exchange Block fire on February 2, 1941. Firefighter Randall began his service with the department as a call man for Engine Company No. 3. He would hold the position of lieutenant in 1944 and 1945 and captain in 1946 and 1947 for the same company. He was honorably discharged from the department on May 5, 1952. (Courtesy of Robert Randall.)

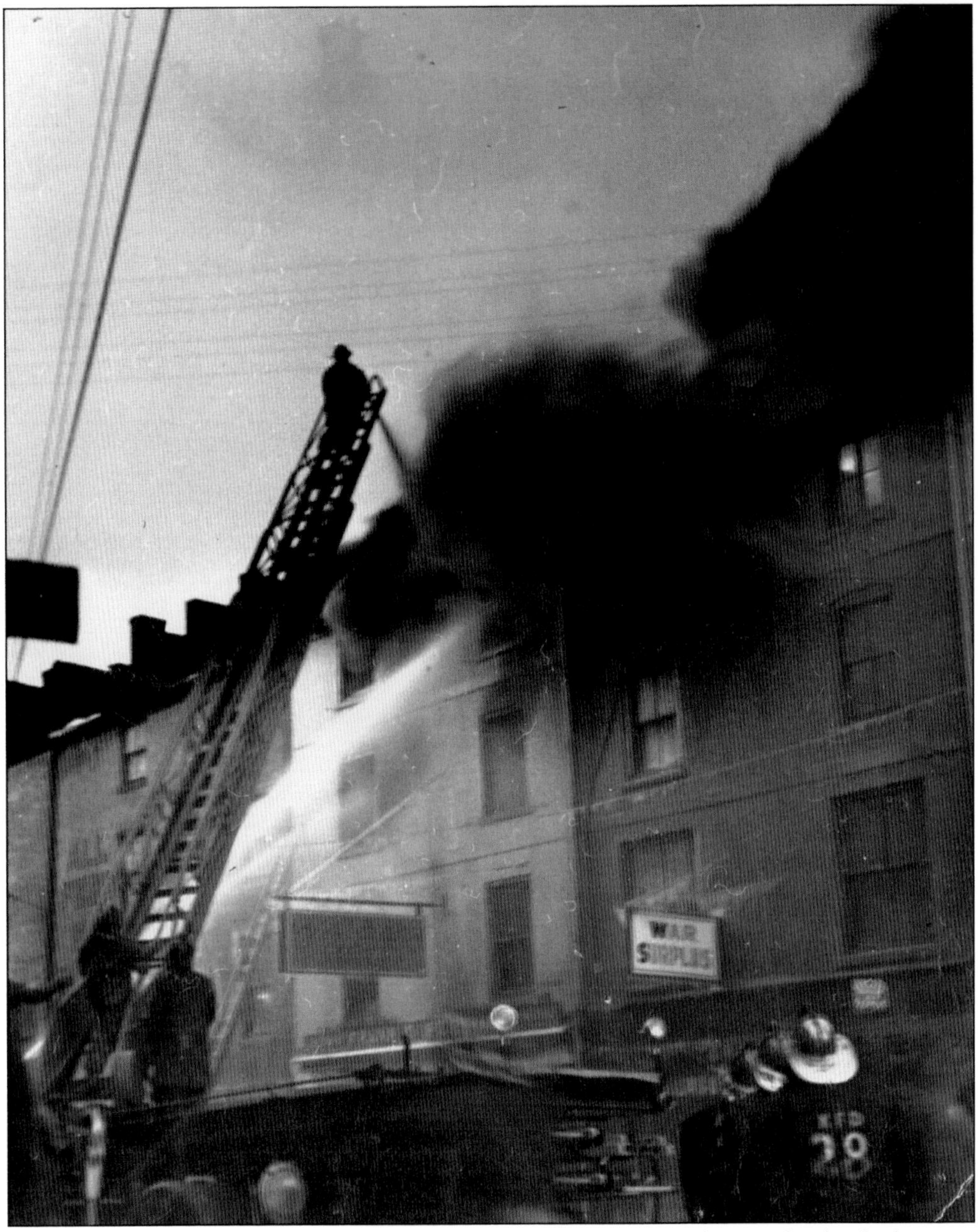

Spectacular billows of thick black smoke could be seen for miles as firemen fought the burning stockpile of rubber contained in the Frank D. Perkins tire recapping plant at 117 Market Street on January 12, 1948. It was reported that at the onset the fire seriously threatened both buildings that flanked the burning structure, but the daring action of the firemen confined the flames and limited outside loss to water damage. It was tough going for the firemen, though, as they fought against the flames and heavy smoke that roared from the building in zero-degree weather. Those who attacked the fire from the rear of the building found themselves sheathed in the frozen spray from the Portsmouth Naval Shipyard's fireboat. The blaze gutted the four-story building of the antiquated business block and destroyed thousands of dollars worth of machinery, supplies, and tires.

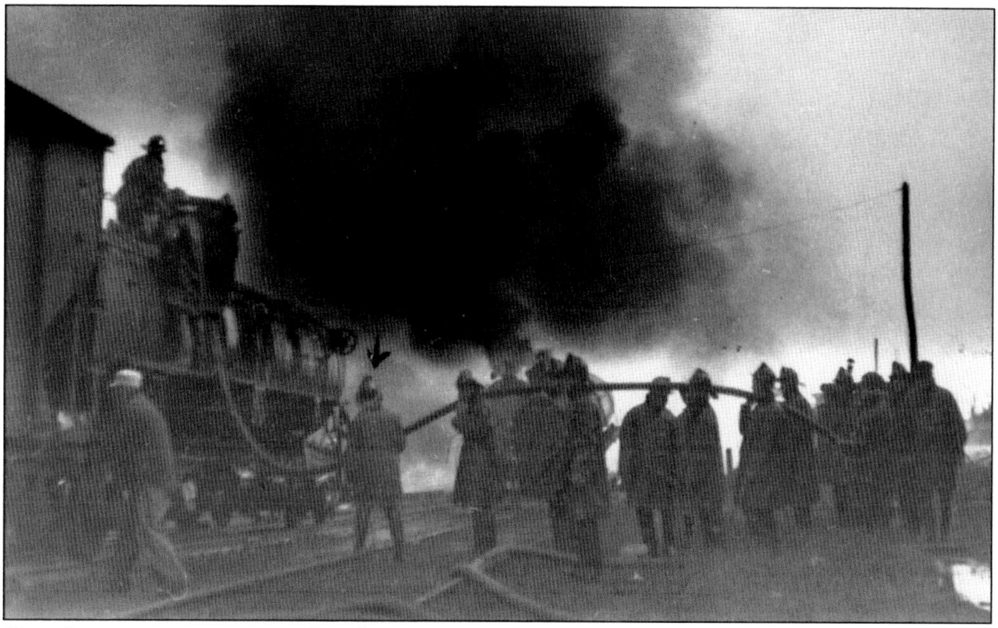

Two alarms were rung in on October 14, 1943, when a Boston-bound passenger train that had just left the Portsmouth depot ran through an open switch and crashed into a gasoline tank car of the Boston and Maine Railway. The collision caused an explosion that destroyed two tank cars, the locomotive of the passenger train, one passenger coach, and three section houses near the tracks. It was reported by a flagman who saw the collision from an engine on a nearby track that the force of the impact caused a sheet of flame to shoot up from the tank car and a massive spray of gasoline to cover the area as far away as 200 feet.

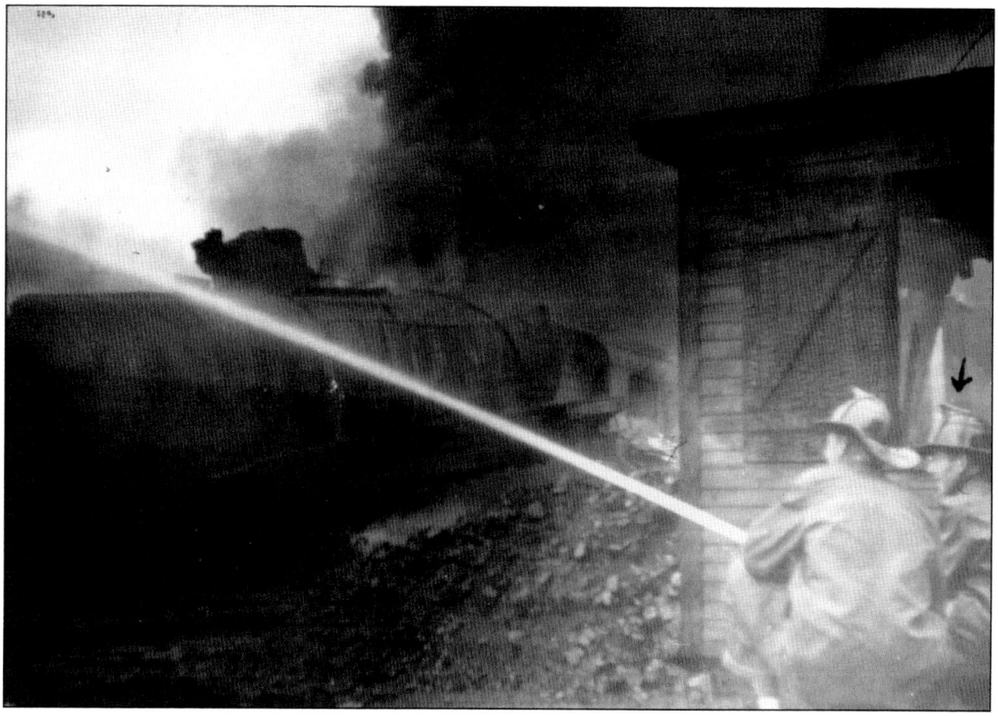

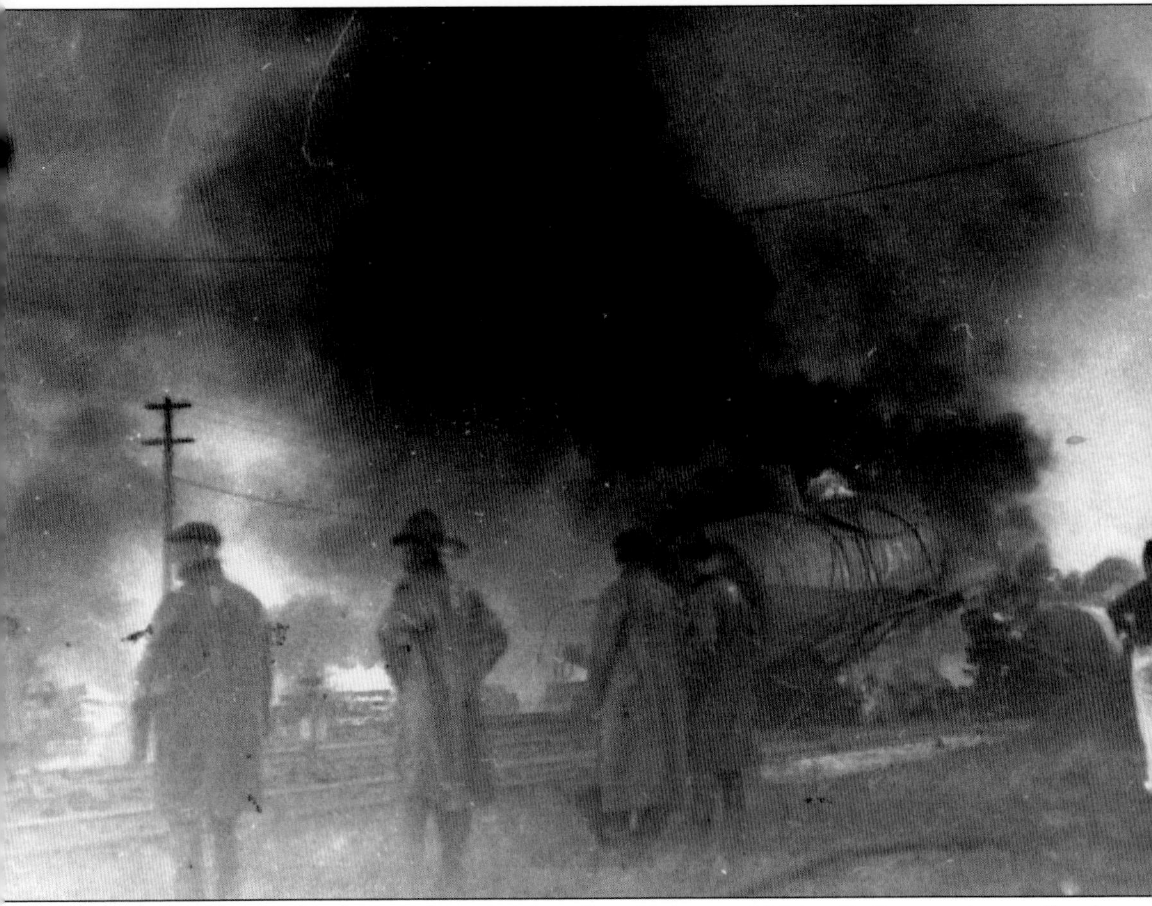

None of the 25 passengers were injured, but three Boston and Maine Railway workers suffered minor burns from the explosion, which was caused when a Boston and Maine Railway gasoline tank car was rammed by the 7:31 a.m. Boston-bound passenger train. Albert Laderbush, an assistant patrol foreman who was in the section house when the collision and explosion occurred just outside the door, was treated at Portsmouth Hospital for first-degree burns. Thomas P. Hannigan, a patrol foreman who was with Laderbush in the shanty, was taken to his home suffering from shock. The engineer of the passenger locomotive and the locomotive fireman were treated at the scene for burns they received when they unsuccessfully attempted to move the locomotive before it was engulfed in flames. Recounted by the flagman who witnessed the crash, "I watched the train coming up the track and was dumbfounded when it swung into the tank car." Fire Chief George T. Cogan estimated the destruction of the cars, properties, and the 20,000 to 30,000 gallons of gasoline would exceed $30,000.

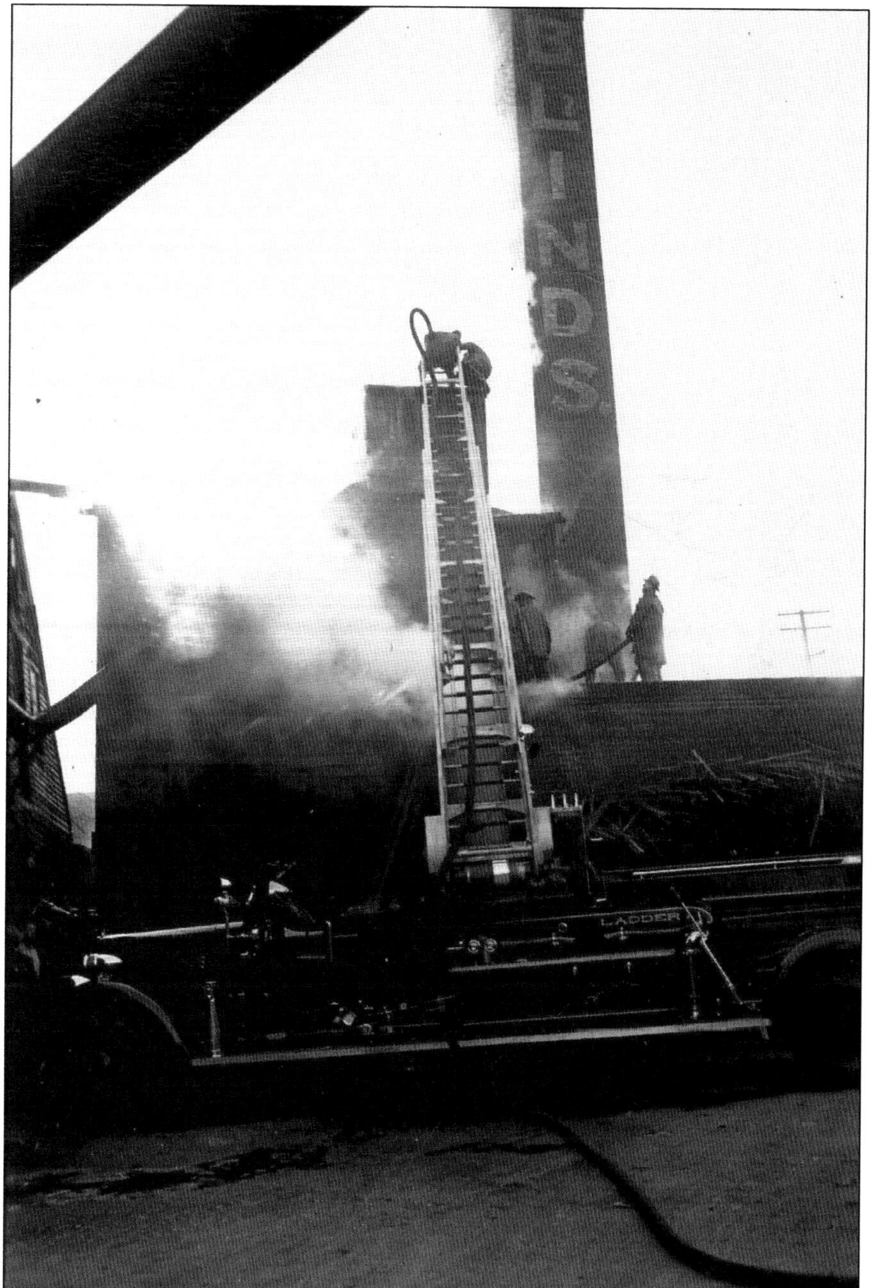

The mill and carpenter shop of the Littlefield Lumber Company at 21 Green Street was severely damaged by a stubborn fire that threatened the entire lumberyard and the nearby homes and businesses. The first alarm was sounded at 1:20 a.m. on June 8, 1951, and the second at just four minutes later. Upon arrival, firefighters found the entire north end of the mill ablaze. Portsmouth crews fought the fire from the Vaughan Street side, while Kittery and Portsmouth Naval Shipyard firemen worked the other side. Handicapped by low water pressure, firemen were required to pump from the Piscataqua River. It was reported that several hundred spectators swarmed into the Vaughn Street area at the sound of the double alarm.

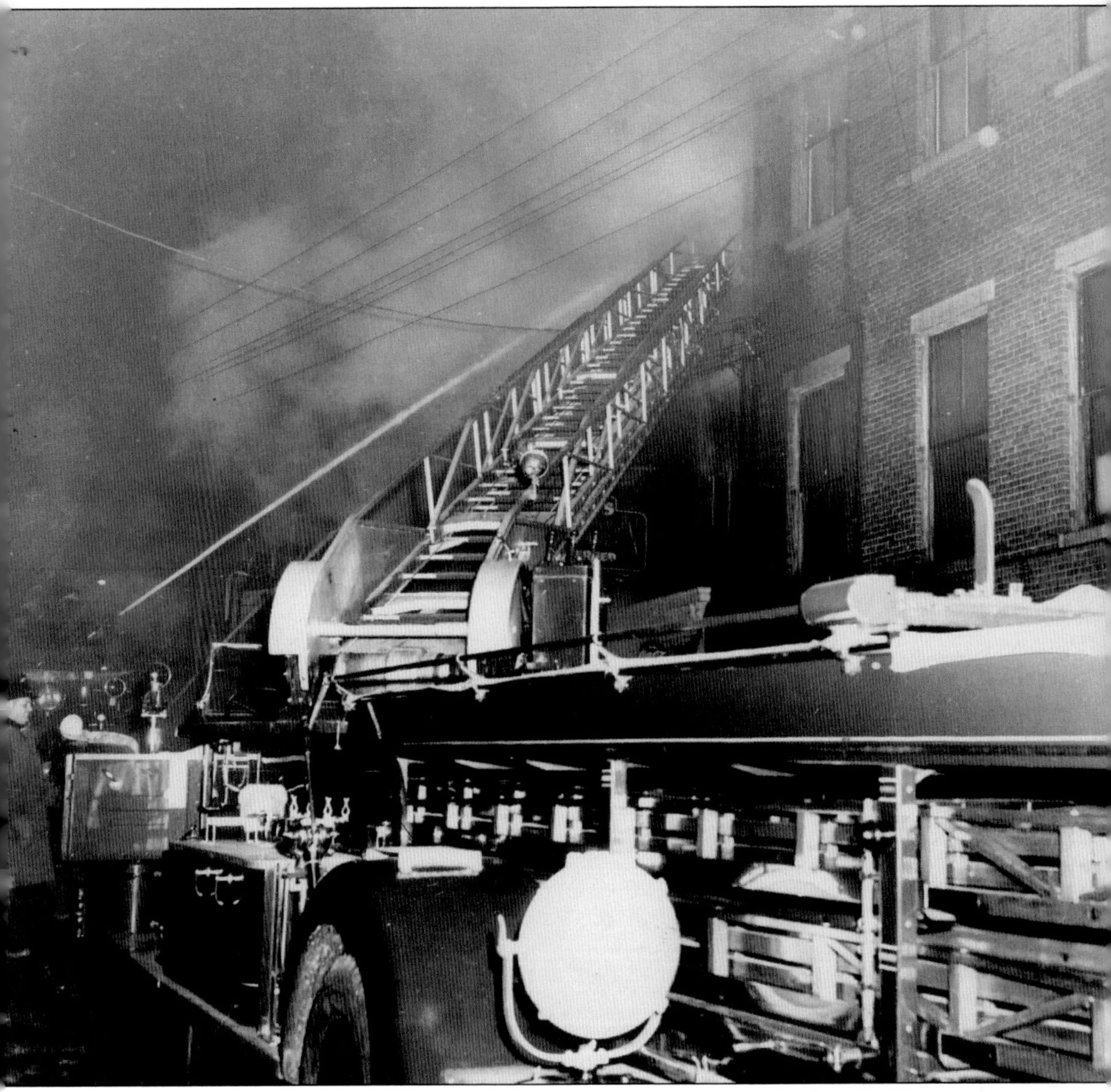

Smoke filled the Kearsarge Hotel and poured out into Congress and Chestnut Streets during a multiple-alarm fire that broke out on January 1, 1956. The blaze was believed to have started around 9:00 p.m. in the kitchen of the Blue Goose Restaurant, which served as the hotel's dinning room. Chief Fredrick R. "Ted" Crompton stated that the flames crept into the walls from the kitchen and spread rapidly to the upper floors. In addition to the restaurant and hotel, Theodore Bosen's hat shop and candy store on Chestnut Street sustained smoke and water damage. During the blaze, close to 700 moviegoers were evacuated from the nearby Civic Theater, albeit a number only after the second alarm was rung in. Firemen who fought the two-alarm blaze in near-zero temperatures and braved clouds of thick black smoke to enter the building were credited with holding down the fire loss and preventing a major conflagration.

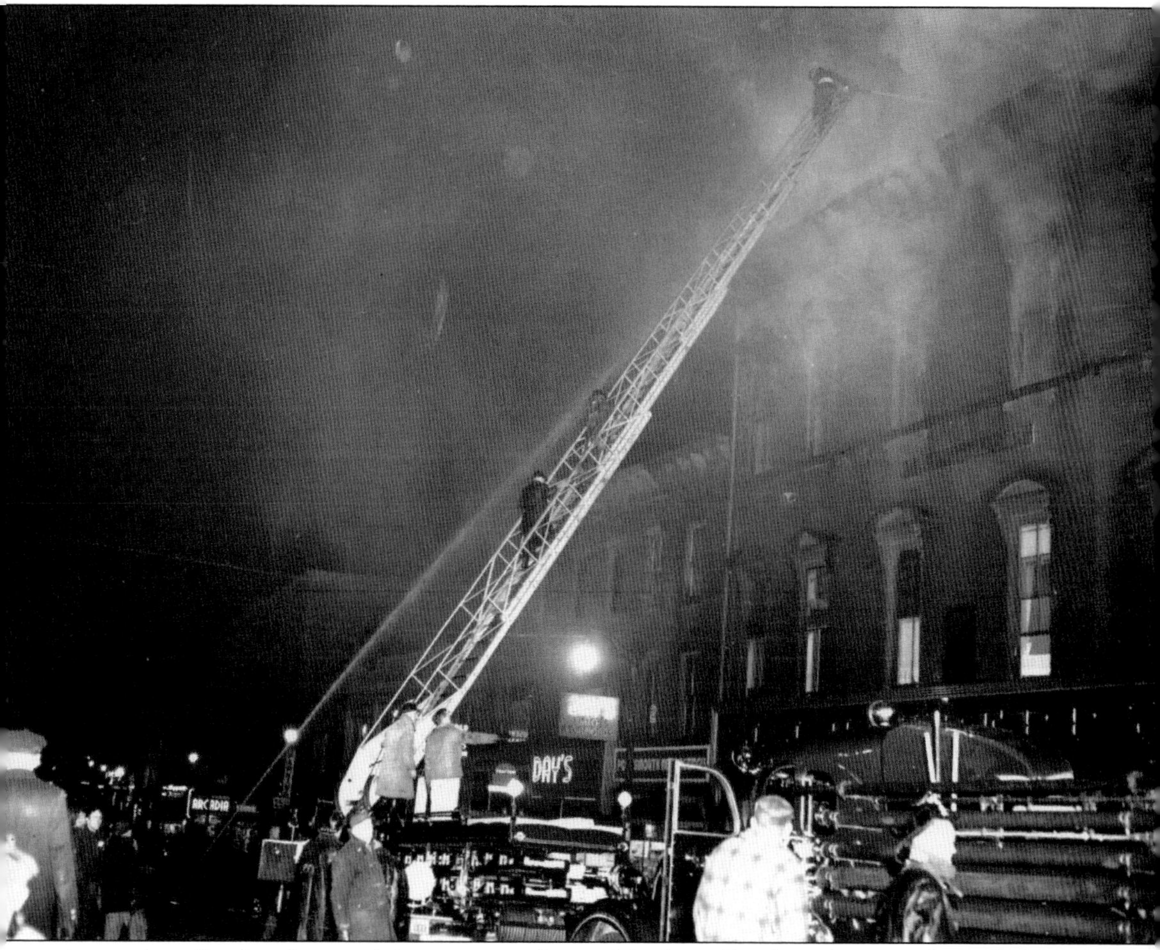

A two-alarm fire that destroyed the upper two floors of the Jarvis Block on Congress Street broke out just before 6:30 p.m. on February 29, 1956. The fire also damaged five nearby business places. Portsmouth and Kittery firemen attempted to battle the blaze on the top floors of the building but were driven out by intense heat and the danger that the slate roof would collapse. The roof would later collapse, falling through to the third floor. Firefighters poured water into the burning structure for more than six hours in an attempt to douse the blaze and were still on the job nearly 12 hours later. Ruined in the fire were the Metropolitan Music School and the Rosale Yosua Dance Studio on the third floor. Damage from smoke and water was extensive throughout the second floor. It was observed that while the still, cold air iced the ladders and roofs the firemen were working from, the lack of a strong wind possibly saved the surrounding buildings from a fate similar to the Jarvis Block. The Congress Block was no stranger to the destructive fury of flames as the 1956 blaze was the third to occur on the site.

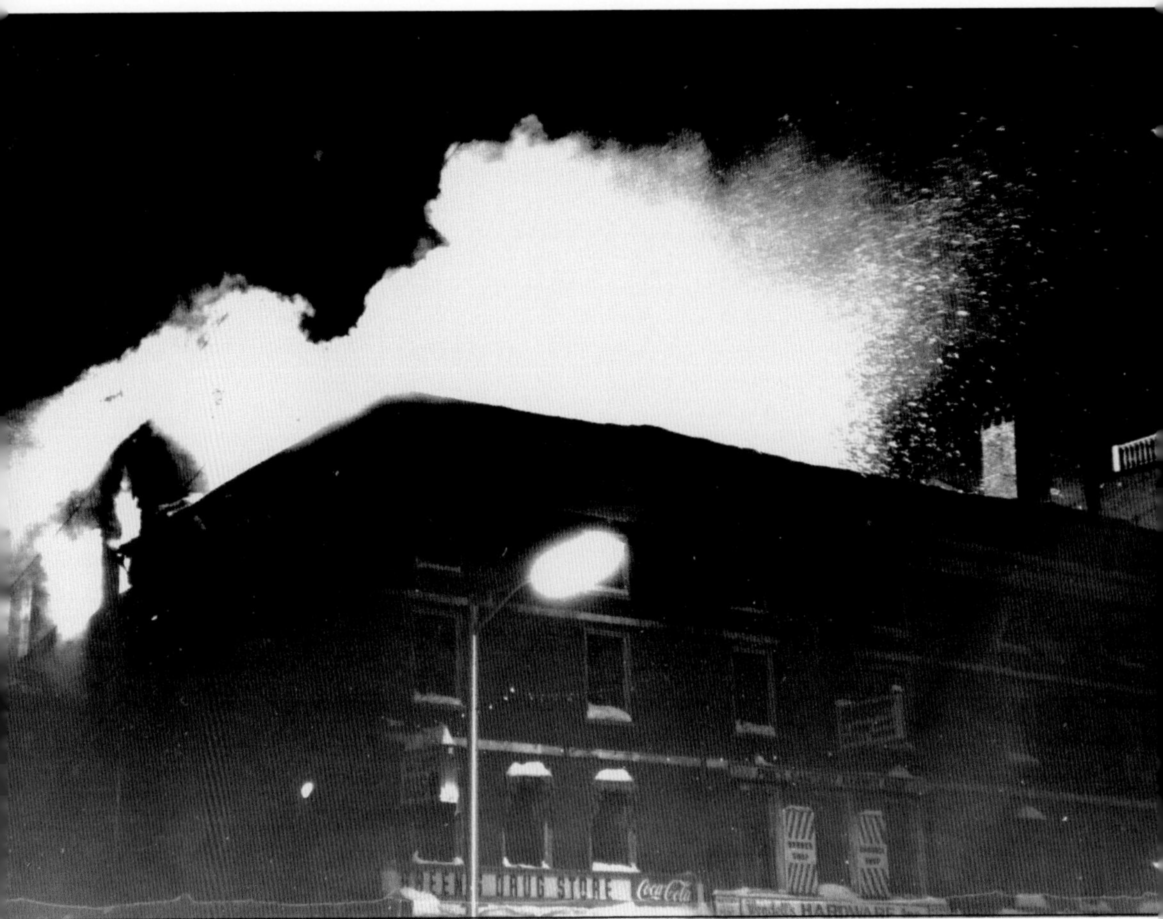

A raging general-alarm blaze, fed by icy 20-mile-per-hour winds, completely destroyed the top two floors of the four-story Pierce building located at High and Ladd Streets on December 14, 1960. It was reported that the fire started on the fourth floor, known as the Moose Hall. The top two floors and roof were completely destroyed as a result of the fire, while the two bottom floors received heavy damage from falling debris, water, and heavy smoke. The lower floors were also occupied by a tailor shop, shoe repair shop, and launderette. More than 100 firefighters from eight towns battled the stubborn blaze for over seven hours. Snow-covered roofs of adjoining buildings prevented sparks from igniting other structures.

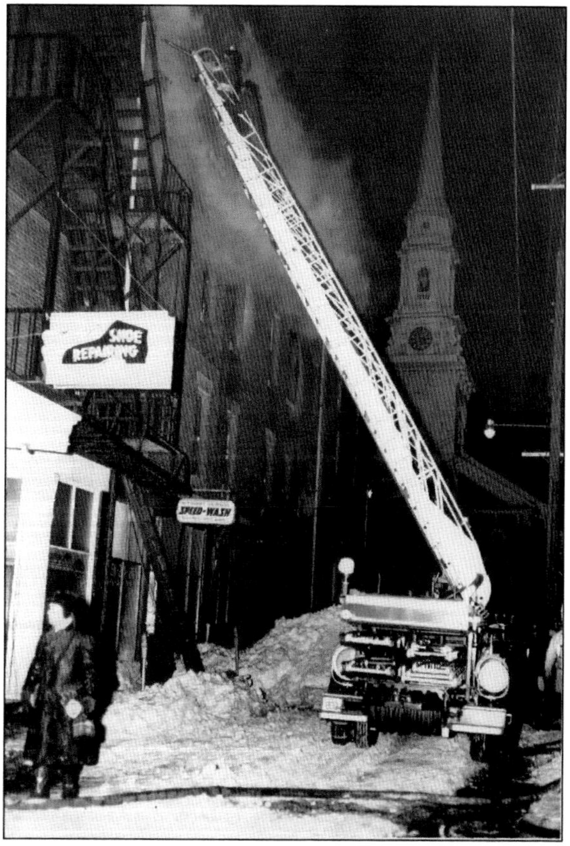

Two Portsmouth patrolmen discovered the December 14, 1960, fire at the Pierce building and rang in the alarm at 1:55 a.m. The second alarm was sounded 10 minutes later. Portsmouth ladder No. 1 was positioned on High Street and the Portsmouth Naval Shipyard fire department's ladder from Ladd Street. Both poured water on the fire into the morning hours. The "all out" came at 9:30 a.m. To the right, the Portsmouth ladder is seen working on High Street with the North Church in the background, and in the image below, a Portsmouth engine is pumping water on Congress Street to crews as the fire gains headway in the upper floors of the Pierce building.

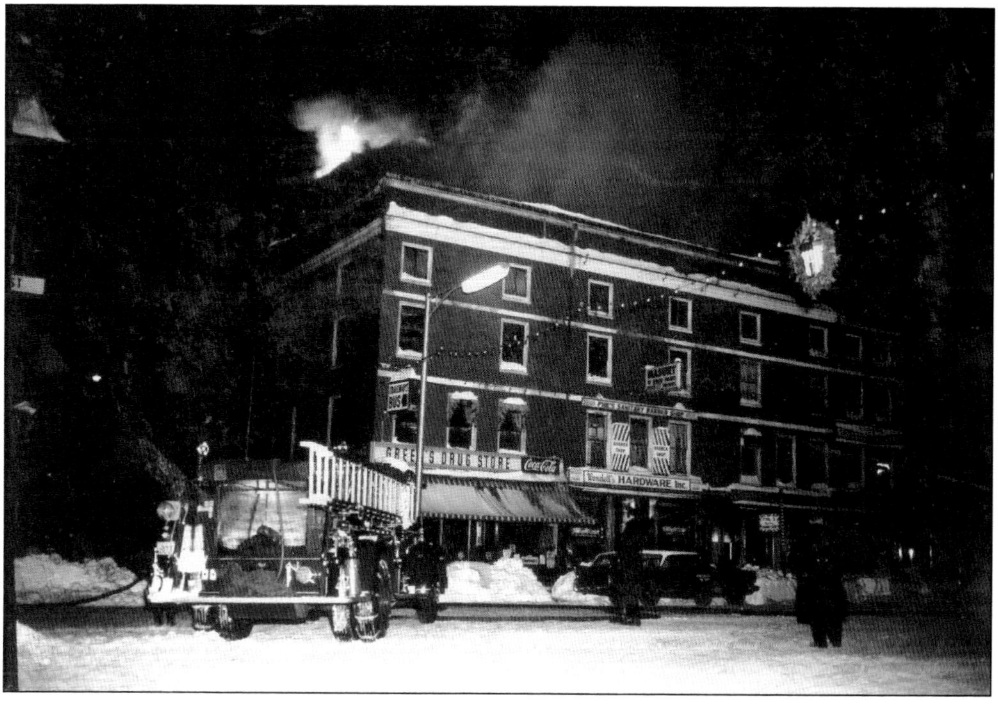

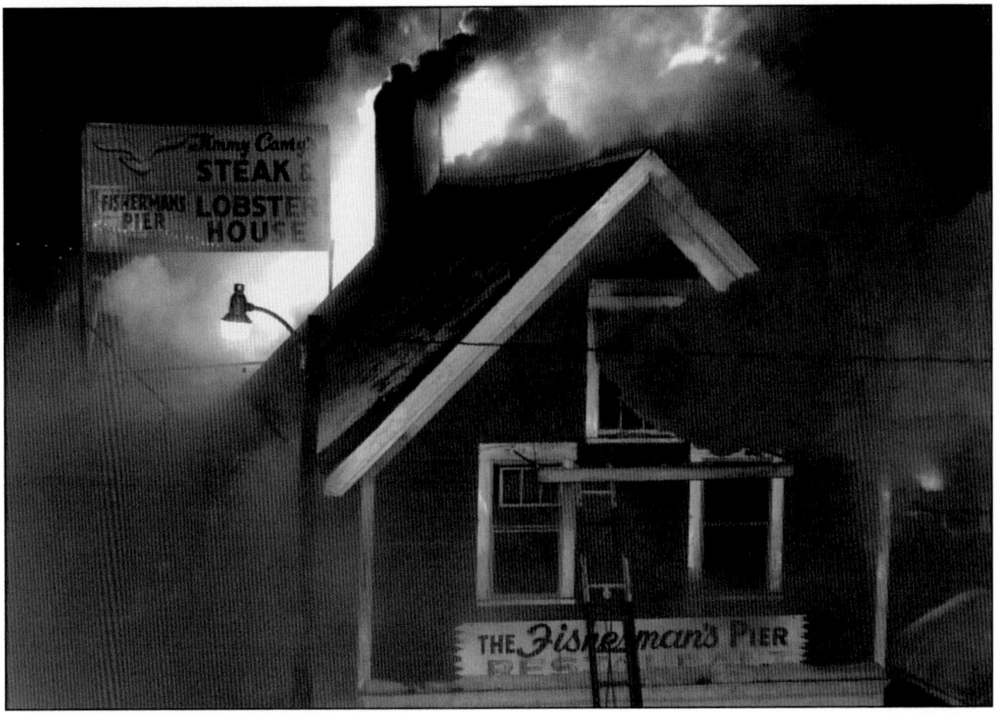

A February 1, 1973, predawn fire engulfed Fisherman's Pier Restaurant on State Street, leaving only smoldering ruins by midday. The first alarm was called in at 5:21 a.m., with the area box alarm ringing in two minutes later. The first firefighting companies were faced with heavy fire consuming the river end of the building with concerns of flying sparks endangering the waterfront area, including Strawbery Banke.

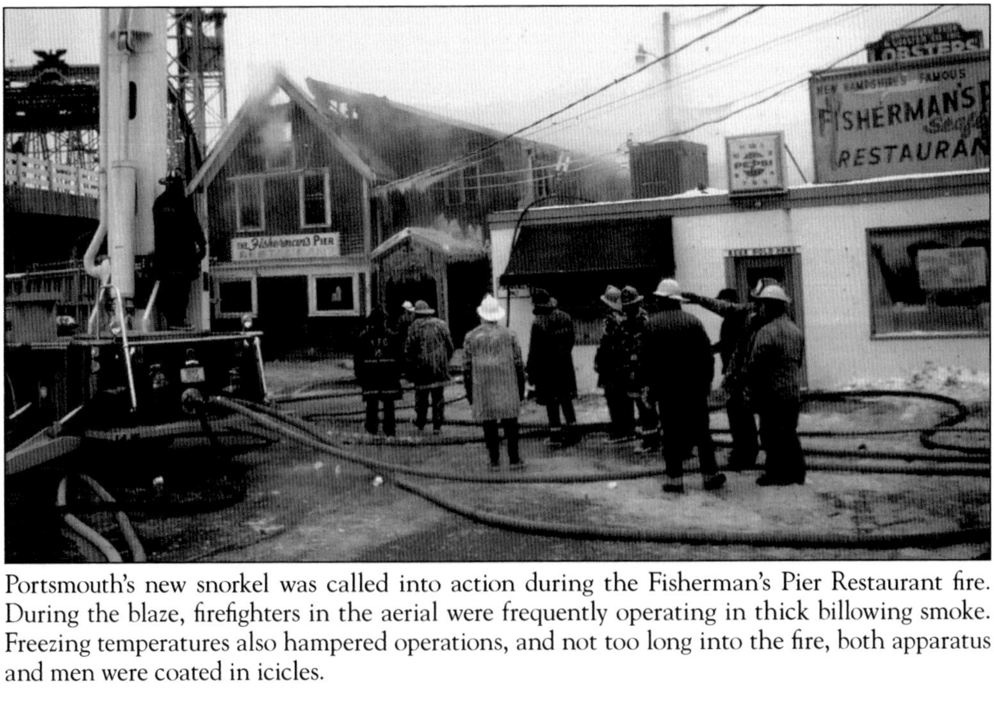

Portsmouth's new snorkel was called into action during the Fisherman's Pier Restaurant fire. During the blaze, firefighters in the aerial were frequently operating in thick billowing smoke. Freezing temperatures also hampered operations, and not too long into the fire, both apparatus and men were coated in icicles.

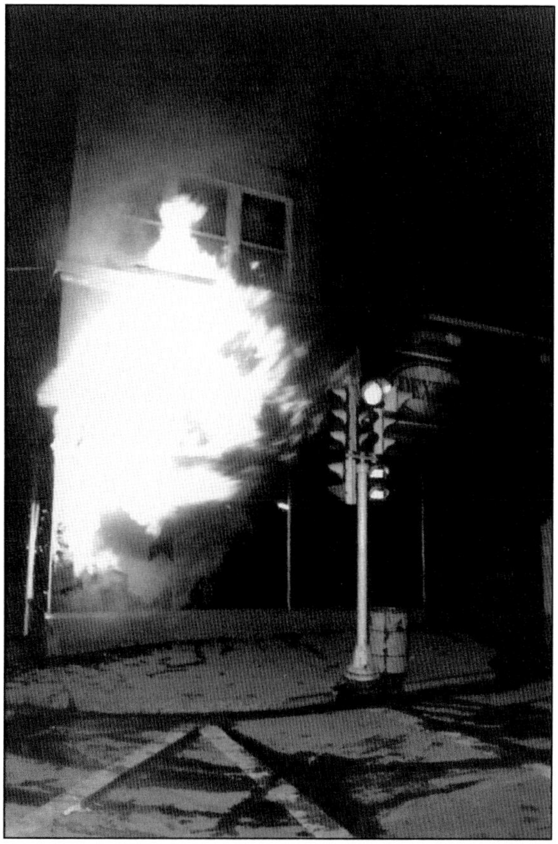

Flames, which at times shot into Congress Street, gutted the Dexter Shoe Factory Outlet on October 7, 1977. The fire was discovered shortly before 4:00 a.m. by Capt. Carl Ward of the Portsmouth Police Department who had just stepped out of the station and saw smoke and a glare in the sky near the downtown area. A box alarm rang in at almost the same time and brought two engines and the ladder to the scene. A second alarm was quickly struck by Lt. John Downs. Due to the intense heat, which blew out all the windows, firefighters were unable to make an interior attack. Most of the efforts were focused on directing hose streams into the storefront and through the windows, as well as spraying down adjoining businesses. (Courtesy of the Portsmouth Herald.)

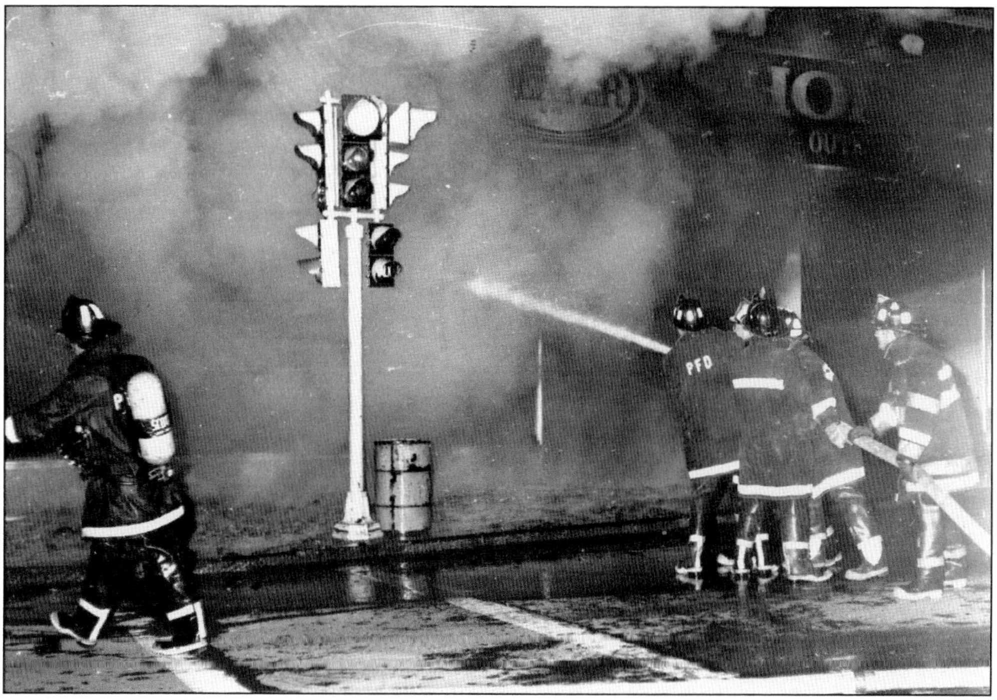

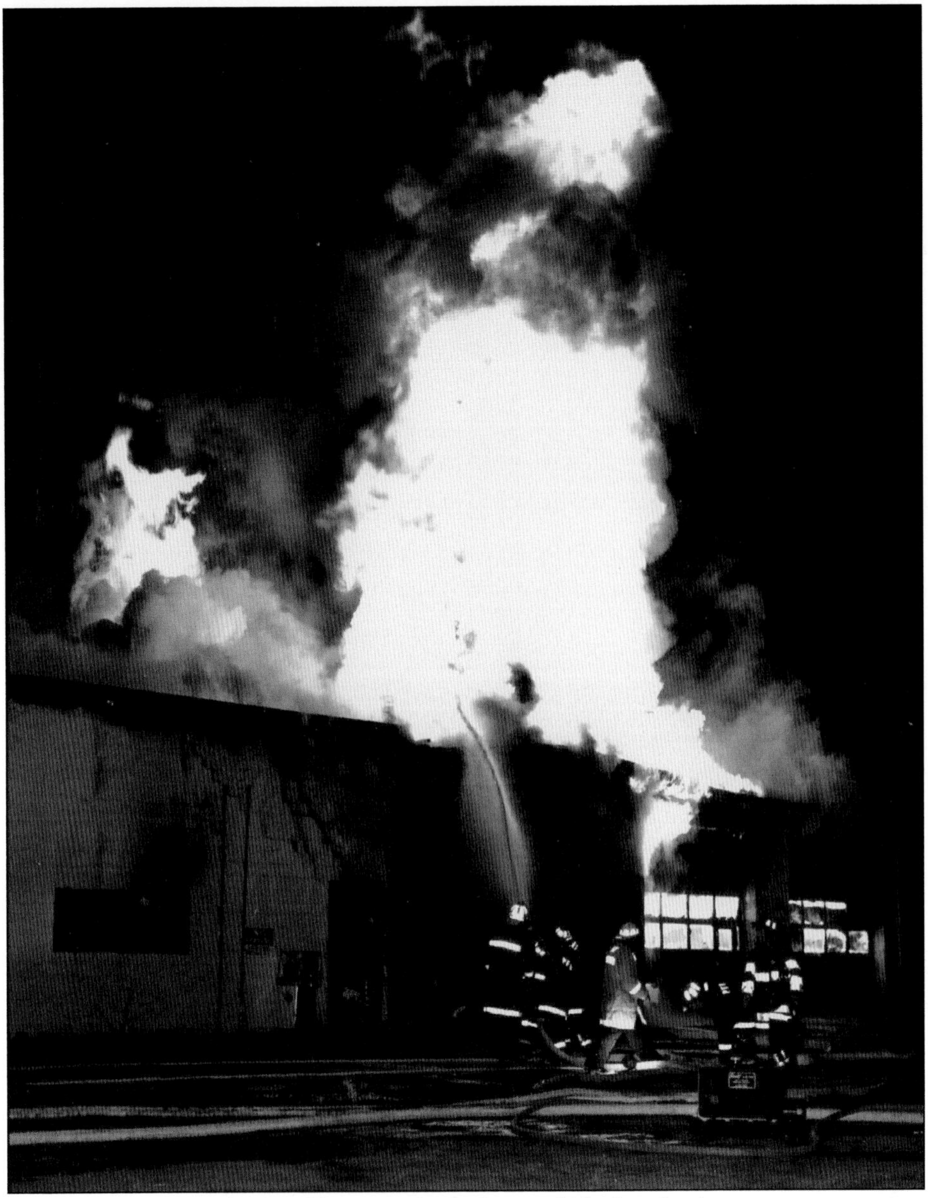

As a result of a suspected safe robbery in which the would-be burglars' attempts at opening the safe with a torch went terribly wrong, the businesses of the Webb Oil Company and Callahan Oil Company at 3000 Lafayette Road were destroyed during this April 14, 1979, second-alarm fire. It was reported that flames reached 100 feet into the sky, and multiple explosions occurred as firefighters struggled to control the blaze. Hoses were turned on the firefighters and apparatus to cool them off as the intensity of the flames engulfed the building. The first alarm rang in at 3:46 a.m., with the second alarm sounding just after 4:00 a.m. The Rye Fire Department was summoned to the scene, while the Greenland, Newington, and Kittery departments covered the city. Columns of black smoke rose into the early morning sky as firefighters fought the blaze that quickly spread throughout the entire building. Two tank trucks that were inside the building, one containing close to 300 gallons of fuel oil, were also destroyed in the fire. By 8:38 a.m., the fire was under control.

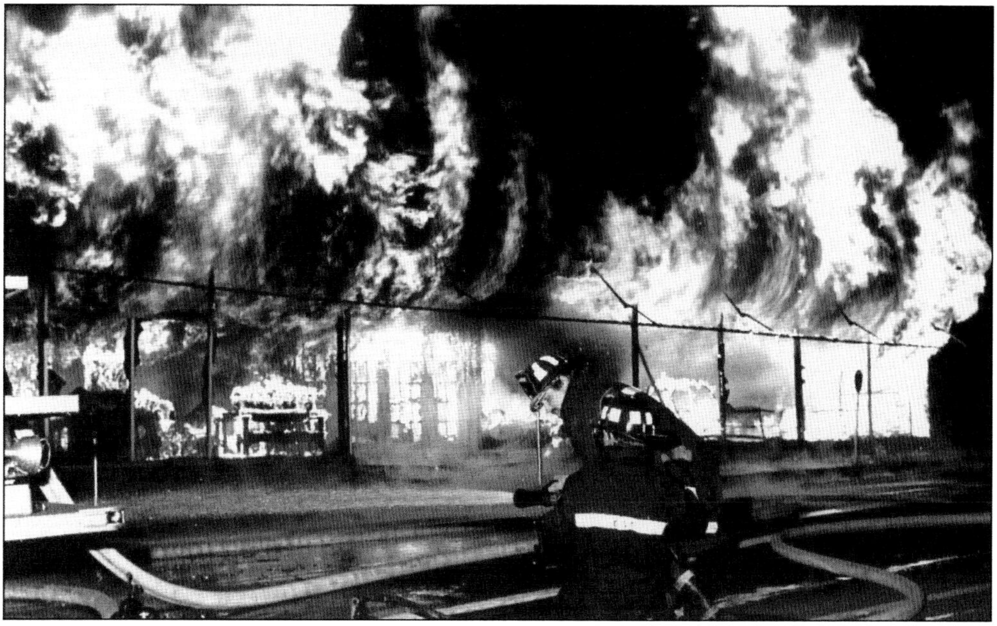

Atherton's Furniture on Islington Street was leveled by a three-alarm fire on Thursday, July 12, 1979. The first arriving engine from the Central Fire Station narrowly escaped damage as a section of the storefront exploded in a ball of fire and debris. Not long after the first alarm was struck, a towering column of fire and smoke could be seen from many parts of the city, attracting a crowd of over 1,000 spectators. By morning, the business was a heap of blackened rubble. Here two unidentified Portsmouth firefighters work among the many lengths of supply and attack hose lines as they prepare for an all-night defensive battle. (Courtesy of the Portsmouth Herald.)

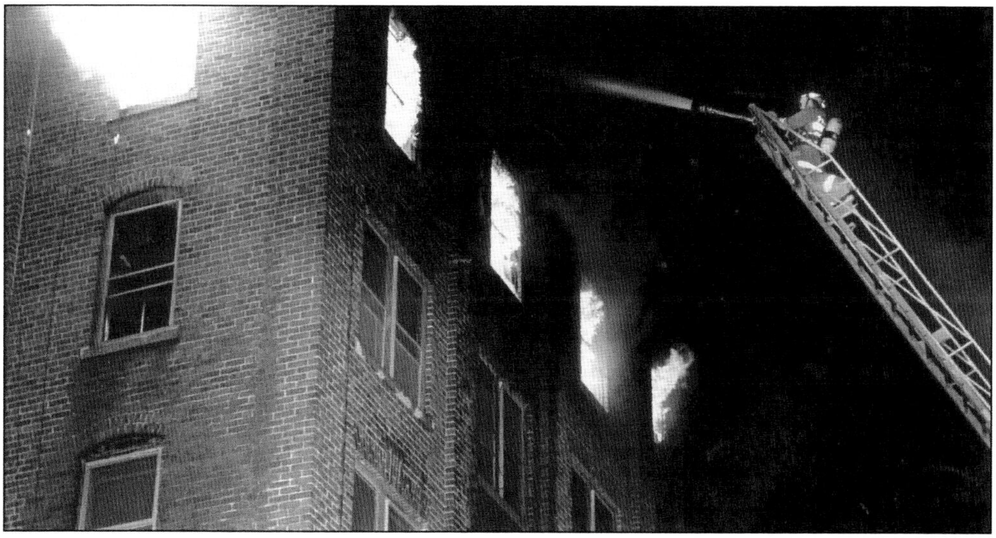

An impressive three-alarm fire raced through the six-story McIntosh building on Congress and Fleet Streets in the late hours of October 13, 1980, destroying the upper two stories and forcing residents to abandon their belongings as they evacuated. It was reported that at one time during the blaze, fire exploded out of numerous windows that then rapidly enveloped the upper floors. Here an unidentified Portsmouth firefighter directs an aerial master stream to the upper floors from ladder No. 1.

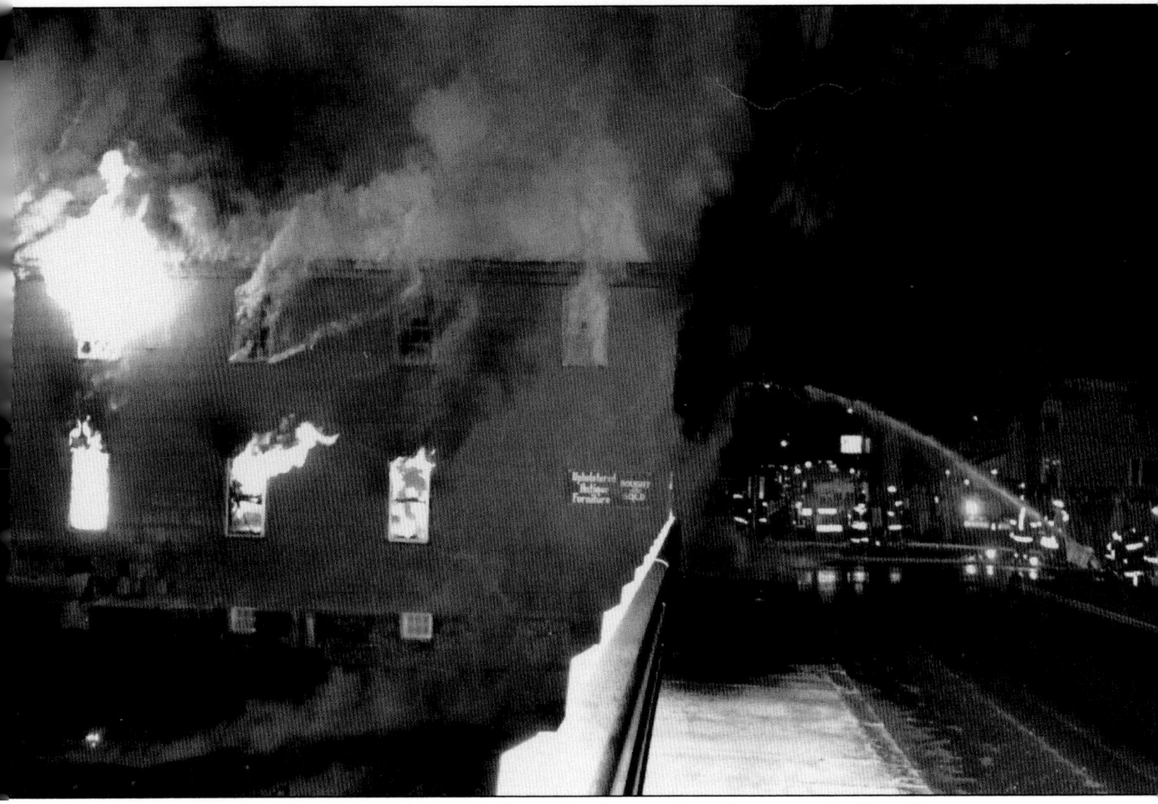

Fire erupts from the windows and through the roof as Portsmouth Fire Department crews strategically position hose lines in a hurried attempt to contain a massive blaze at the Model Upholstery building at 230 Maplewood Avenue. The first box rang in at 8:03 p.m. on Sunday, February 3, 1980. Shortly thereafter, the first truck wheeled up to the scene to find heavy fire and smoke conditions, prompting a special call for additional manpower. According to Chief Paul Long, the fire was going for quite some time before the department got the call, and upon arrival, the structure was fully involved. It was reported that the repeated ringing of empty air packs could be heard in the cold, clear night as billows of dense smoke poured into the night air.

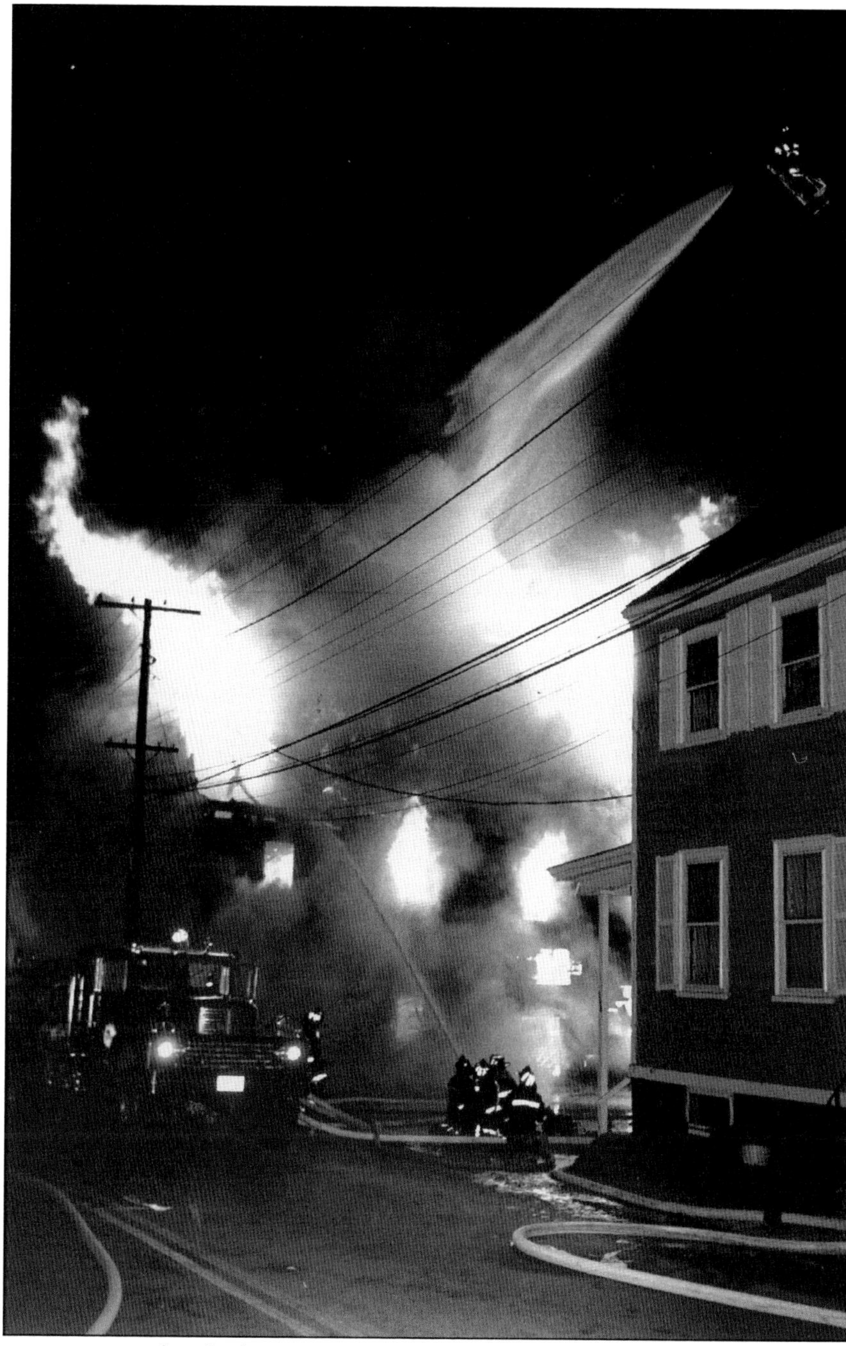

Ladder operations and multiple master streams could not save the Model Upholstery building on February 3, 1980. The building was being used to store furniture and for reupholstering work. It contained foam rubber, plastics, and refinishing supplies that accelerated the growth and intensity of the blaze. Smoke poured from the eaves and then billowed into the night sky. Eventually fire would shoot from the right rear of the building, and shortly after, a large portion of building came crashing down when both the rear and right sides failed. That was followed by the front wall falling into the street. The fire was declared under control at 9:25 p.m.

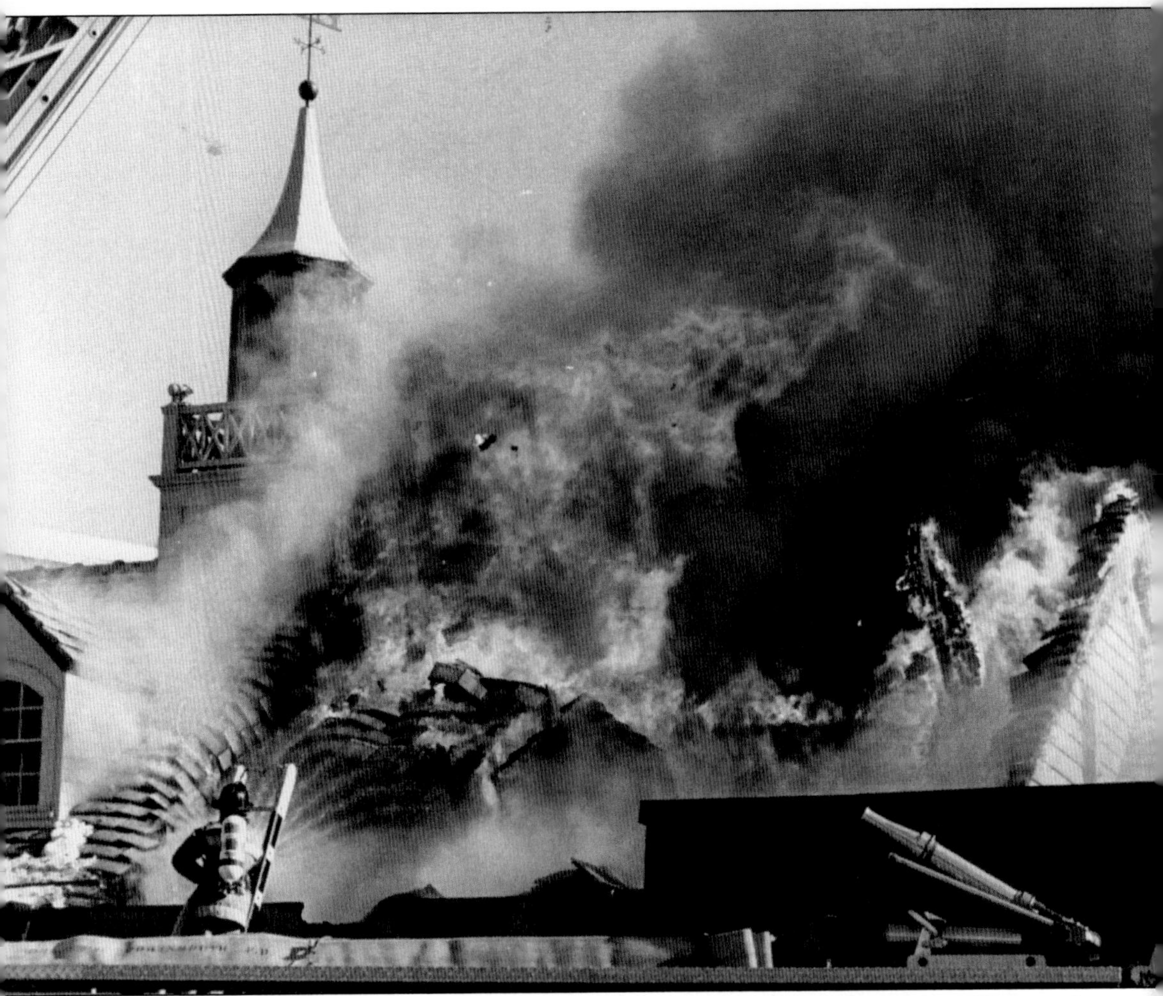

At 11:24 a.m. on January 20, 1981, the department responded to a fast-moving blaze at the Howard Johnson's restaurant located just off the Portsmouth traffic circle. According to Fire Chief Paul Long, the interior of the attic was fully involved when firefighters arrived, and they were driven back by intense heat and smoke. The wind also drove the fire back down through the roof, which collapsed along with the cupola. The fire was contained to the area of the restaurant with minor damage to the new motel addition. With an initial alarm of two engines and a ladder, a second alarm was sounded, bringing in off-duty personnel and mutual aid. An additional problem faced by fire crews was that there were no hydrants immediately accessible, requiring 1,200 to 1,500 feet of large-diameter hose to be laid from Woodbury Avenue.

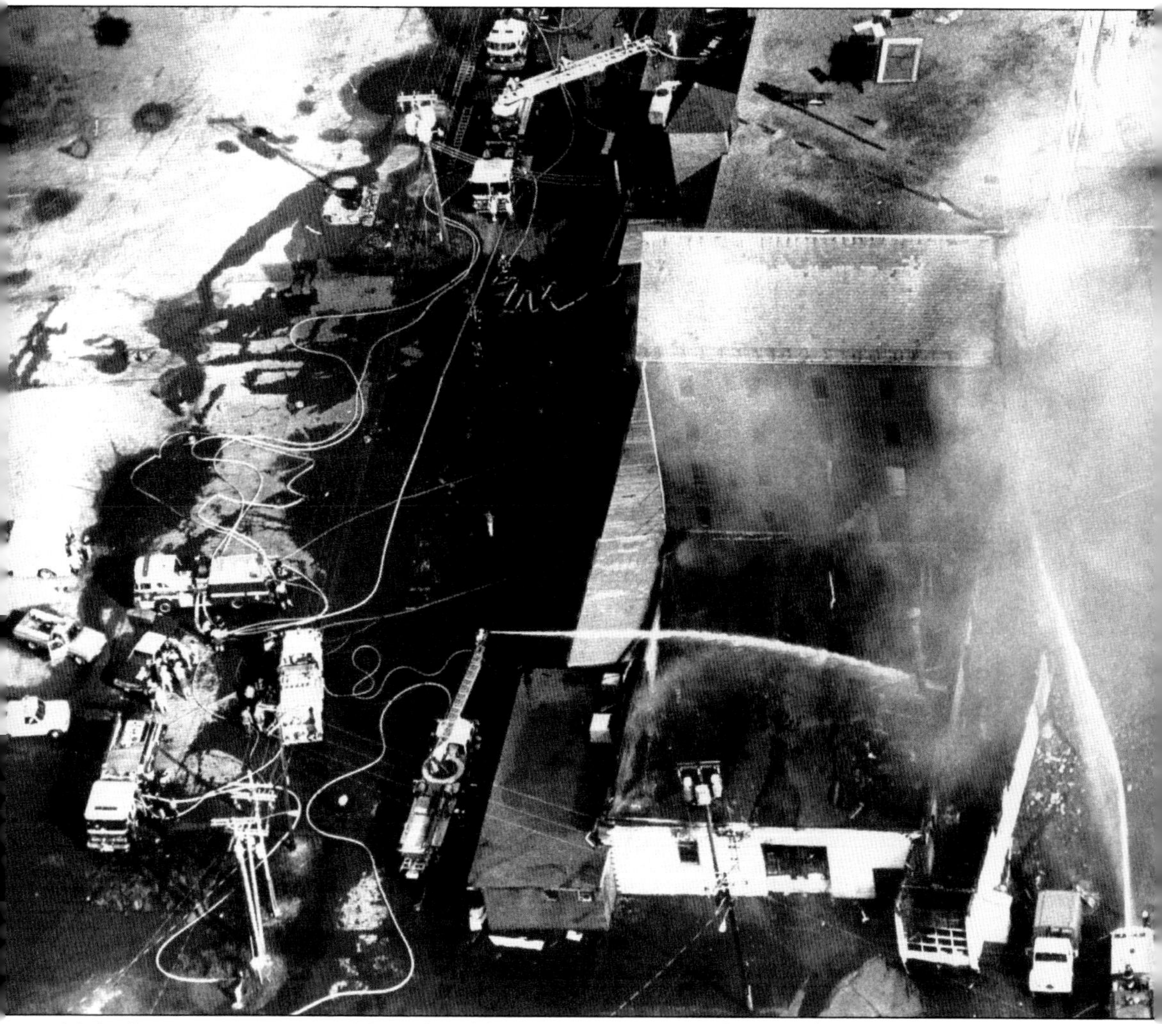

Multiple master streams and large hand lines are in action in this aerial view of a three-alarm blaze that ripped through the Schultz Meat Specialties Plant located at 698 Islington Street on Wednesday, June 13, 1990. More than 100 firefighters from 15 area fire departments were engaged in containing the blaze at the former Frank Jones building. This was not the first time that the building, also known as New Hampshire Provisions Co., was heavily damaged by fire. The first devastating fire at the New Hampshire Provisions Co. occurred on October 21, 1957. That fire broke out around 10:25 p.m. and was visible for miles, drawing hundreds of spectators. Portsmouth and mutual aid fire companies laid nearly a mile of hose and poured close to 780,000 gallons of water in attempts to extinguish the multiple-alarm blaze. And it was on a blistering July 15 afternoon in 1983 that a massive blaze erupted, keeping firefighters busy for more than five hours.

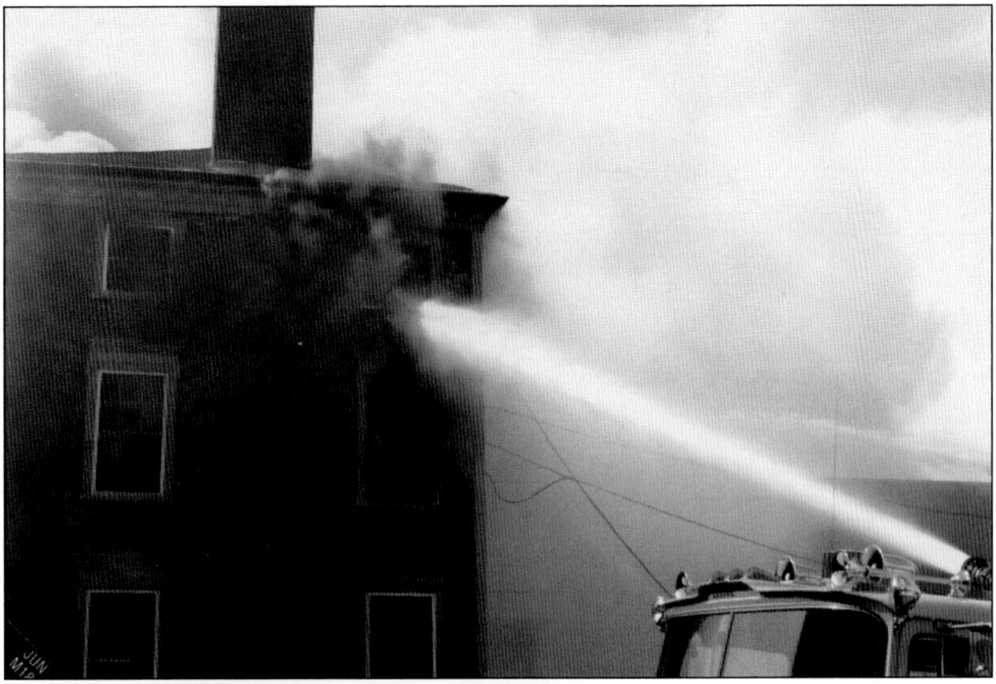

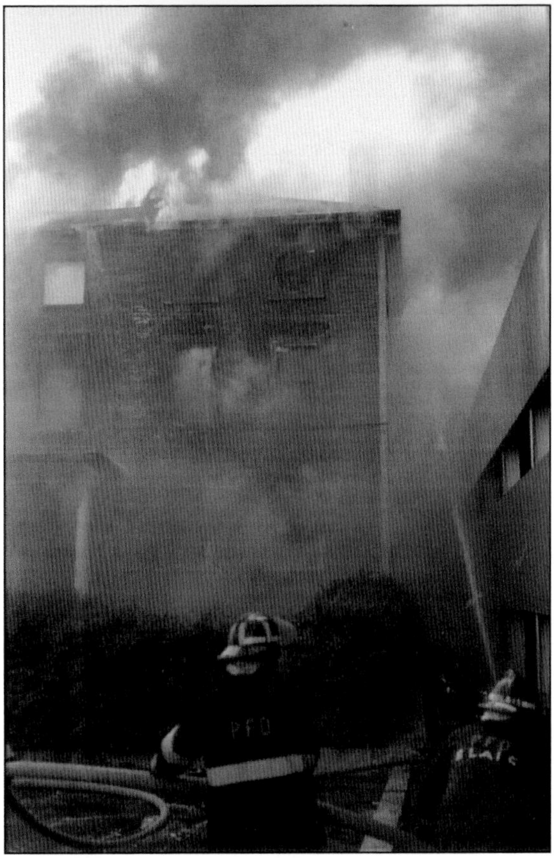

These two photographs were taken on June 10, 1986, as firefighters battled a challenging three-alarm blaze that severely damaged a 185-year-old vacant structure at 44 High Street. In an attempt to get ahead of the fire and protect exposures, the standpipes of the neighboring parking garage were employed by firefighters. The adjacent Newberry's Department Store was scorched in places, but there was no damage to the High-Hanover parking garage only a few feet away. Four weeks earlier, the structure suffered significant interior damage when an arsonist touched off a fire at the property.

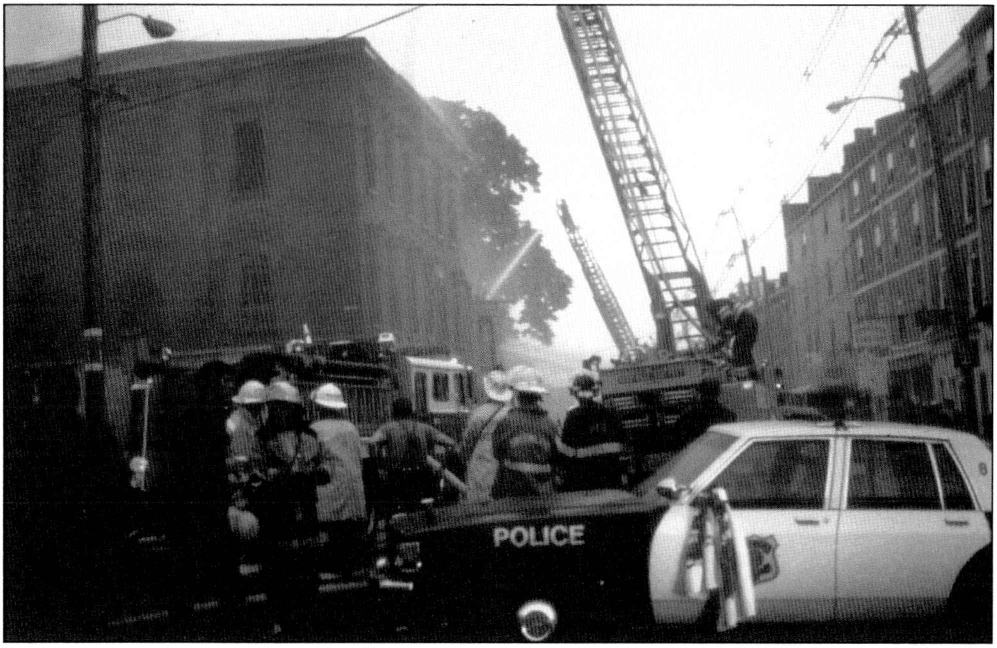

On July 23, 1989, a five-alarm fire broke out at 1:45 a.m. at the Standard Plumbing and Heating storage building located at 126–128 Market Street. Upon arrival, firefighters found both the second and third floors heavily involved with fire rapidly extending to the adjacent 116–118 addresses. Multiple ladders and pumpers are still in operation in this photograph taken on the morning after.

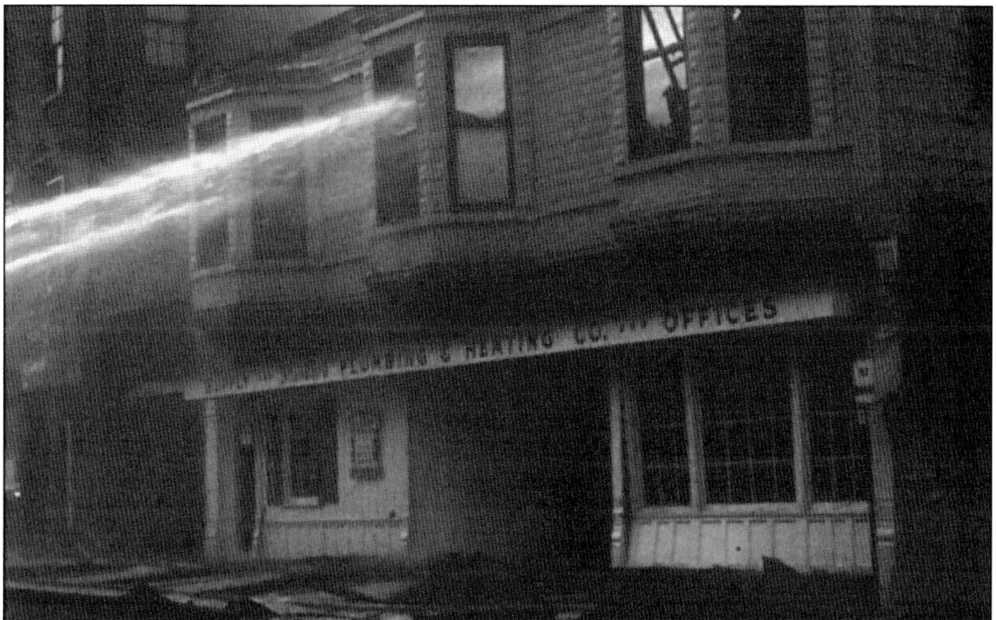

Fire crews direct hose lines into the remnants of the Standard Plumbing and Heating building in this photograph taken on July 23, 1989. At the height of the fire, crews pumped water from the Piscataqua River to supplement the large volumes required to contain and extinguish the blaze.

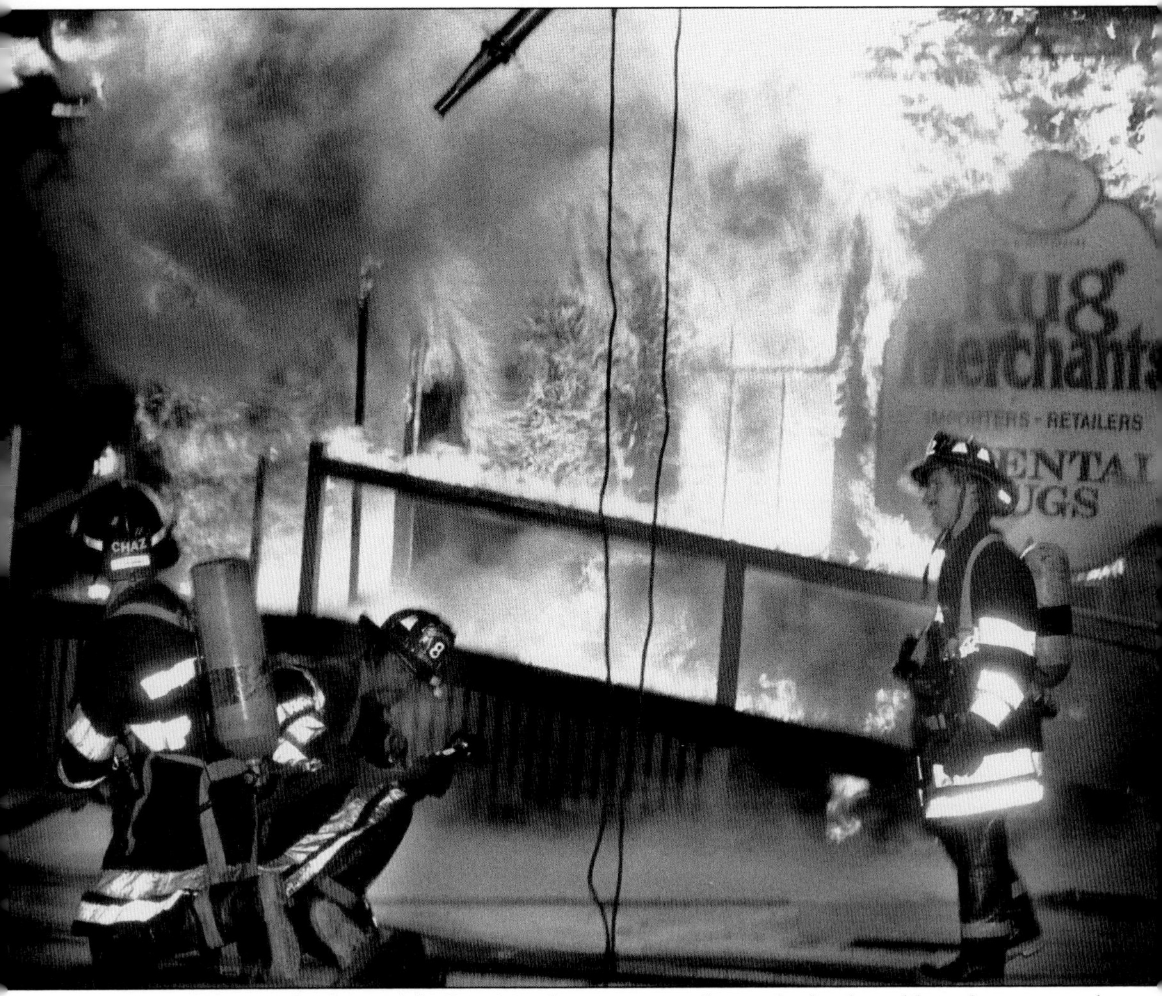

Flames blow out the front of the Rug Merchants store at this multiple-alarm blaze that occurred on November 6, 1989, at the Deer Street Depot located at the corner of Deer and Bridge Streets. Eventually the fire would roar through the entire building and cause the roof to collapse. Seen here positioning a hose line in an attempt to contain the blaze are, from left to right, firefighters Charlie Cormier, Richard B. Duddy, and Bernie Marvin. The first alarm was sounded for a fire at the 165 Deer Street location at 6:38 p.m. and took close to three hours to control. (Courtesy of the Portsmouth Herald.)

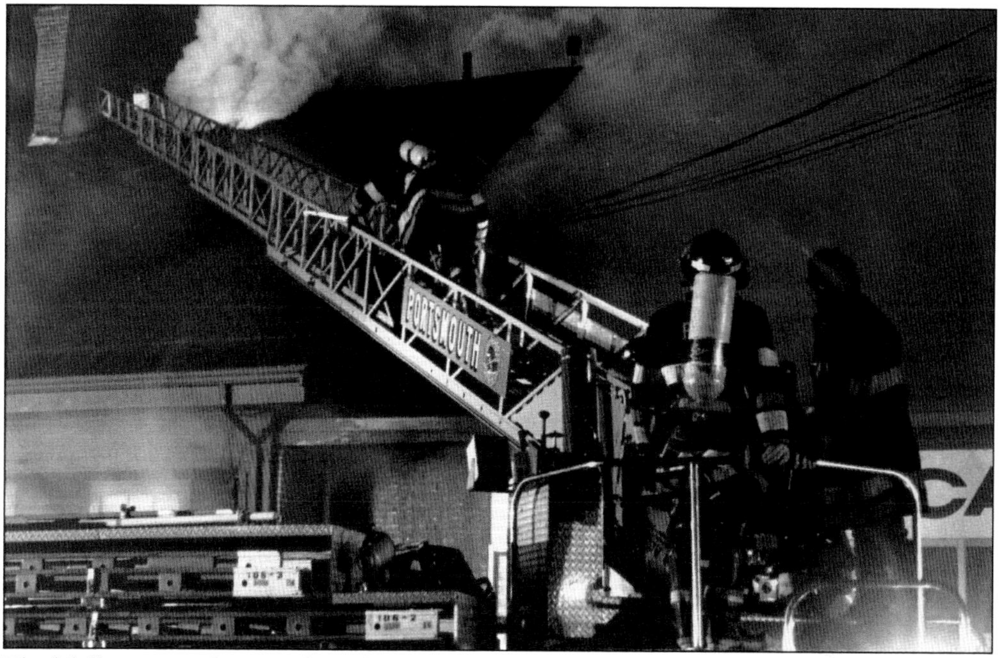

Thick billowing smoke pours from the roof of the Deer Street Depot following ventilation work by firefighters Frank Ott, Kenneth Smith, and Timothy Collins. A passerby noticed smoke coming from the former Boston and Maine Railway freight shed building, and the fire department was called around 6:25 p.m. Within an hour, flames shot up through the roof after fire engulfed the enclosed space above the ceilings. (Courtesy of the Portsmouth Herald.)

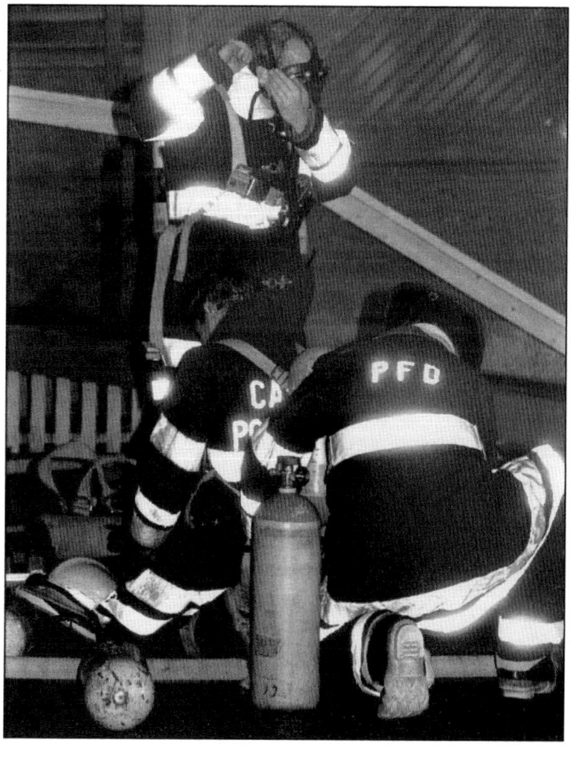

Firefighter Timothy Collins is seen here fitting his air mask as Capt. Edward Tully gets his bottle changed out by an unidentified Portsmouth call firefighter at the Deer Street Depot blaze. The fire on November 6, 1989, destroyed the Rug Merchants and former Softworks stores at the Deer Street Depot along with causing extensive damage to two adjacent stores. (Courtesy of the Portsmouth Herald.)

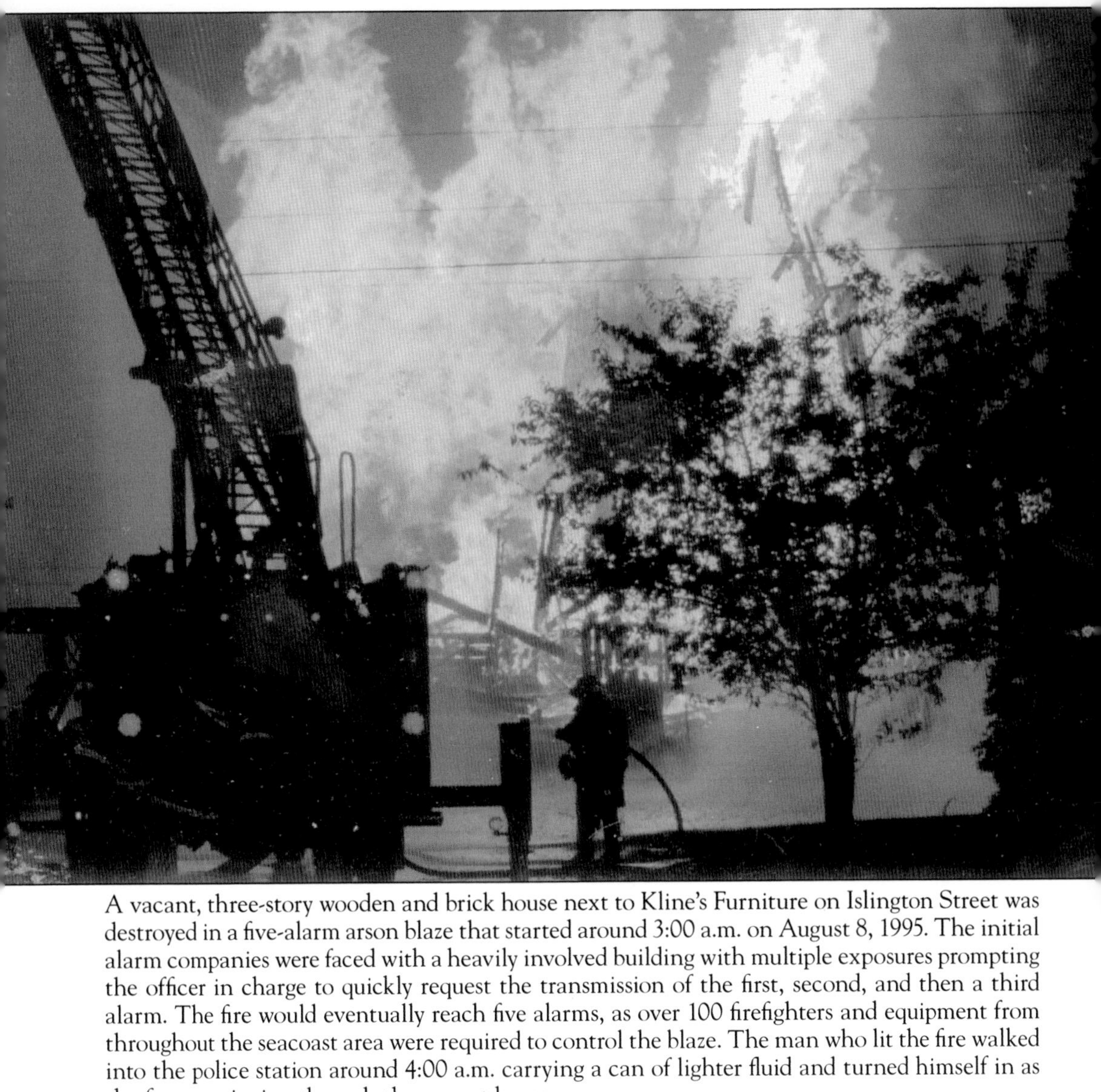

A vacant, three-story wooden and brick house next to Kline's Furniture on Islington Street was destroyed in a five-alarm arson blaze that started around 3:00 a.m. on August 8, 1995. The initial alarm companies were faced with a heavily involved building with multiple exposures prompting the officer in charge to quickly request the transmission of the first, second, and then a third alarm. The fire would eventually reach five alarms, as over 100 firefighters and equipment from throughout the seacoast area were required to control the blaze. The man who lit the fire walked into the police station around 4:00 a.m. carrying a can of lighter fluid and turned himself in as the fire was ripping through the vacant home.

Seven

Beyond the Alarm

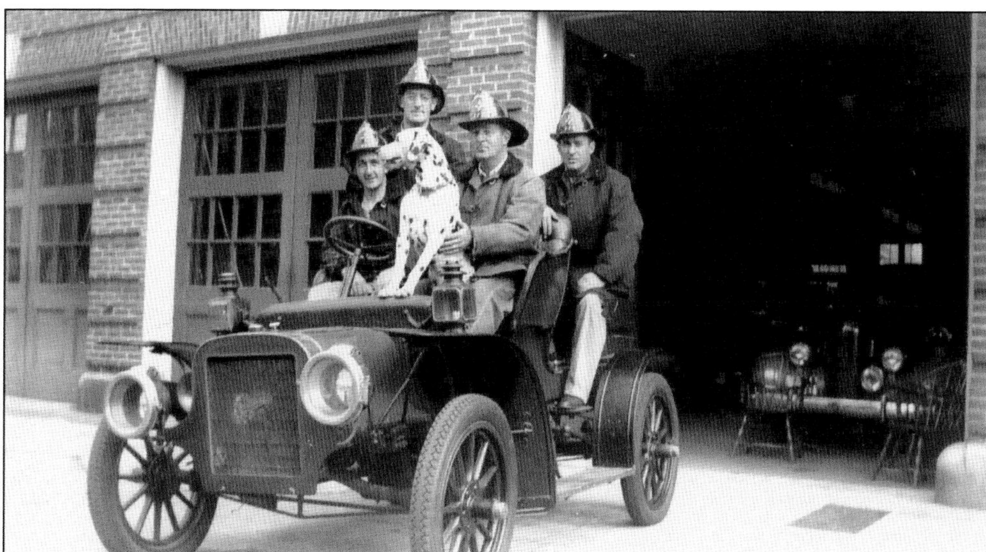

Chief George T. Cogan's dalmatian looks forward to a ride in the parade in this image taken sometime around 1945. The dog was given to the chief by a local Portsmouth physician and was promptly named Doc. Along for the trip are, from left to right, (front seat) firefighters Ernest Weeks and Franklin Davis; (back seat) Tom Kelly and John "Bill" Connors. (Courtesy of the Connors family.)

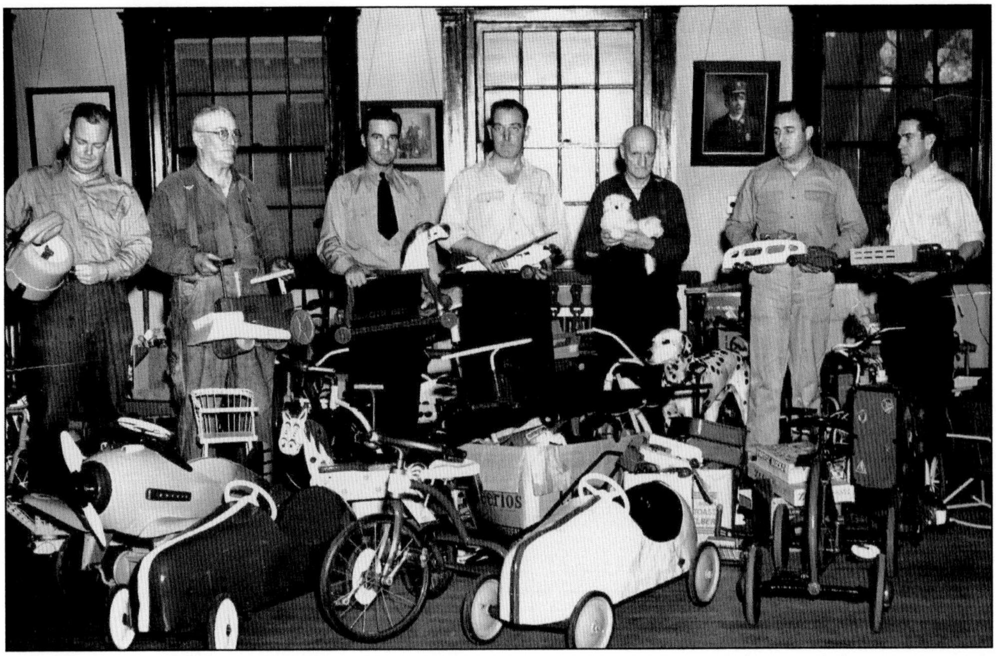

Started by Chief William F. Woods in 1925, each Christmas the firefighters of the Portsmouth Fire Department would repair and recondition donated toys for distribution to needy and underprivileged children. Carrying on the tradition in this photograph taken at the Central Fire Station during the late 1940s are, from left to right, Bob Murray, Arthur Cox, Eliot Staples, Tom Kelly, Chief George T. Cogan, Bill Connors, and Virgil Bagley.

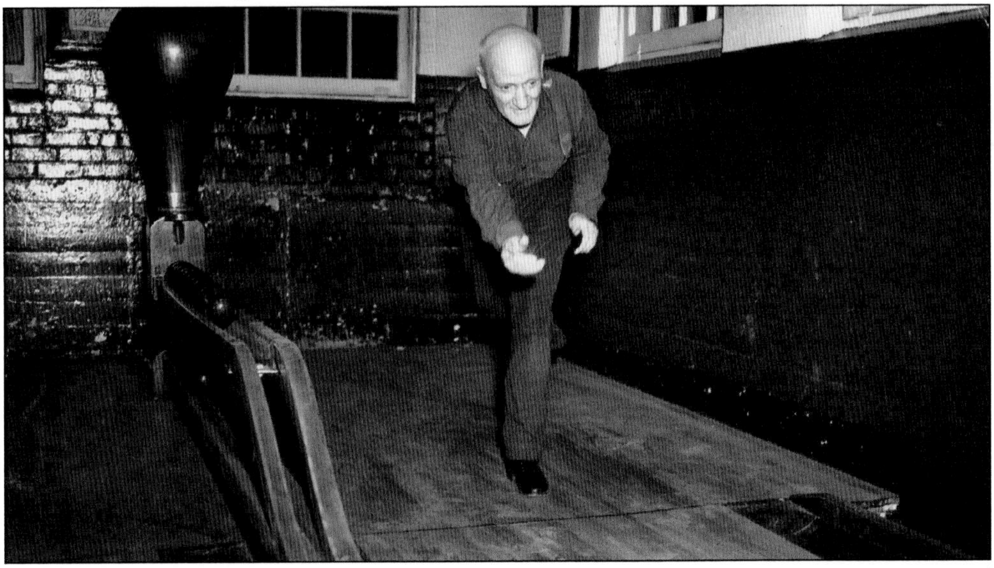

Reported to be the only firehouse in the country to feature a bowling alley, the new Central Fire Station built in 1919 had a working two-lane wooden alley located in the basement. The bowling alley was used by both firefighters and bowling leagues until it fell in to disrepair in the early 1950s. This image of Chief Cogan bowling was taken sometime during the 1940s. Note that the large ornament between the lanes is an air chamber from one of Portsmouth's steam fire engines.

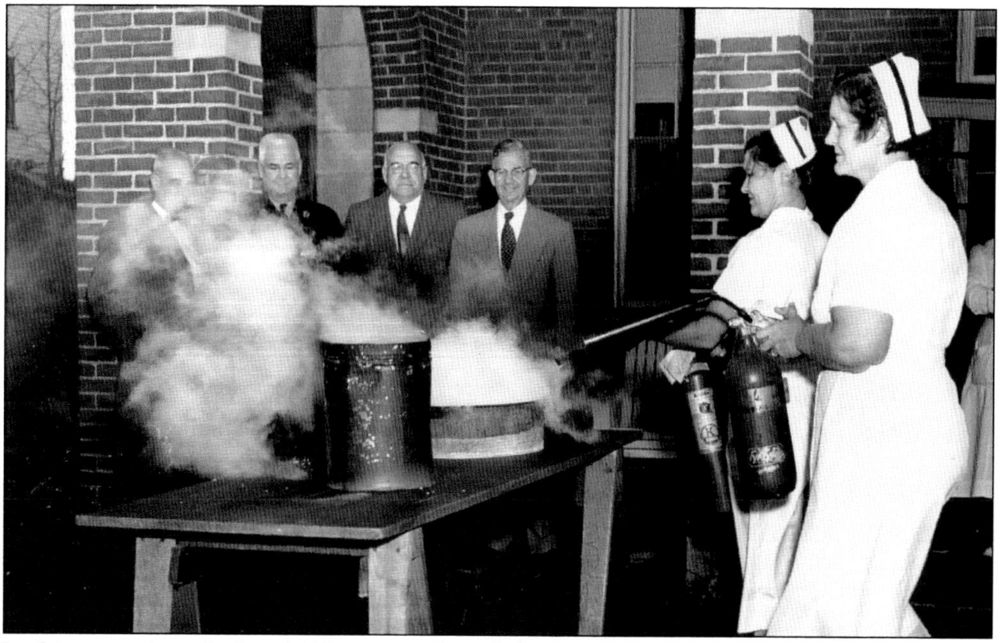

In this image from the late 1940s, two unidentified nurses demonstrate the operation of portable fire extinguishers during a fire-prevention training session at Portsmouth Hospital. Looking on are, from left to right, Fire Marshal Aubrey Robinson, Deputy Fire Chief Fredrick R. "Ted" Crompton, Stimson Taylor from the board of fire control, and Mayor Andrew Jarvis.

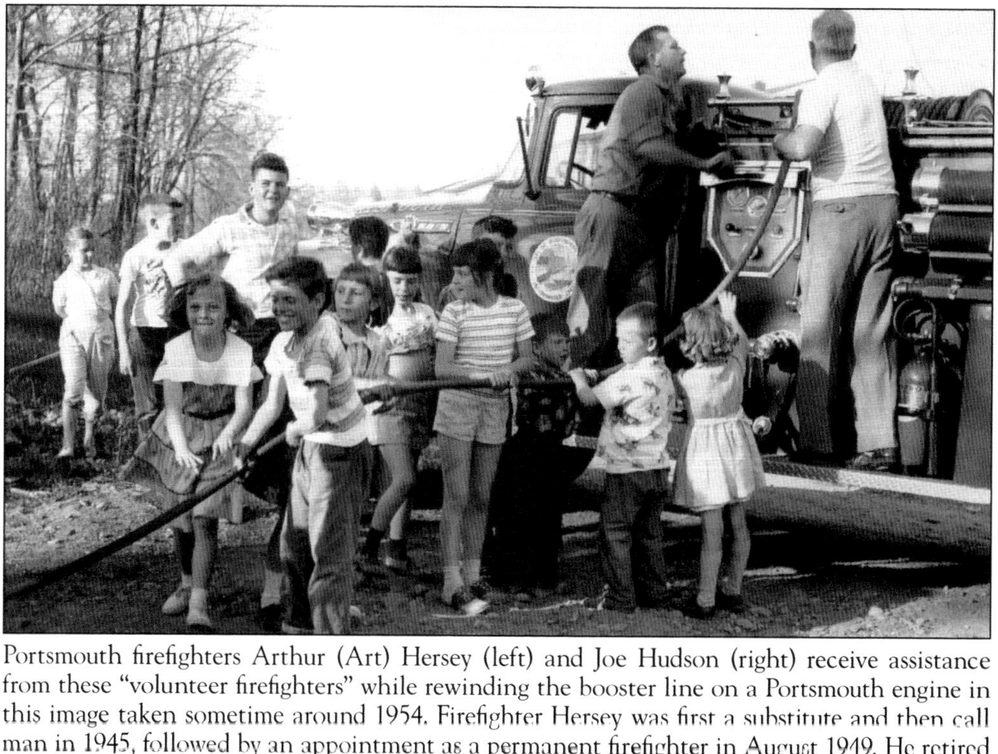

Portsmouth firefighters Arthur (Art) Hersey (left) and Joe Hudson (right) receive assistance from these "volunteer firefighters" while rewinding the booster line on a Portsmouth engine in this image taken sometime around 1954. Firefighter Hersey was first a substitute and then call man in 1945, followed by an appointment as a permanent firefighter in August 1949. He retired from the department in 1970.

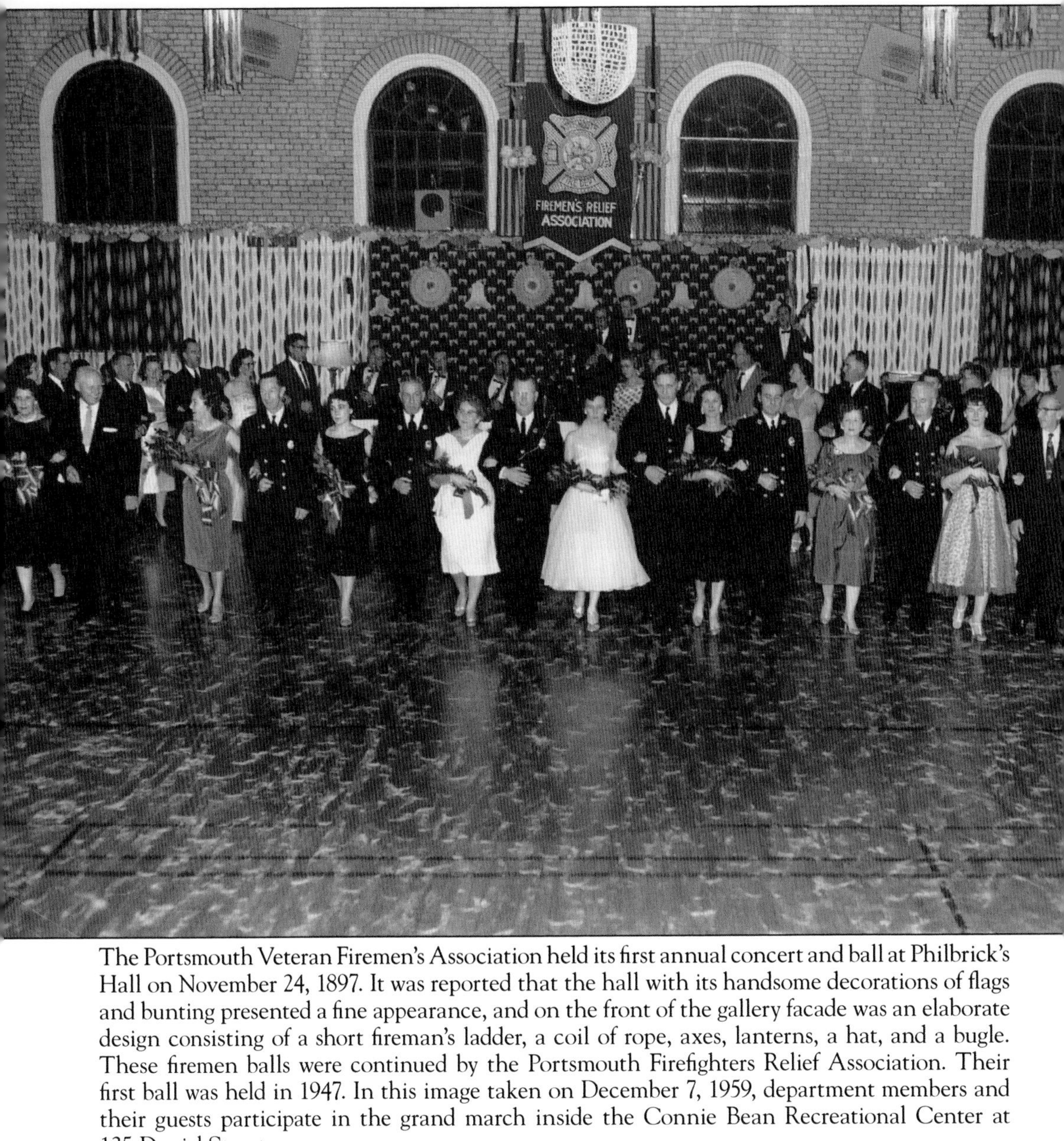

The Portsmouth Veteran Firemen's Association held its first annual concert and ball at Philbrick's Hall on November 24, 1897. It was reported that the hall with its handsome decorations of flags and bunting presented a fine appearance, and on the front of the gallery facade was an elaborate design consisting of a short fireman's ladder, a coil of rope, axes, lanterns, a hat, and a bugle. These firemen balls were continued by the Portsmouth Firefighters Relief Association. Their first ball was held in 1947. In this image taken on December 7, 1959, department members and their guests participate in the grand march inside the Connie Bean Recreational Center at 135 Daniel Street.

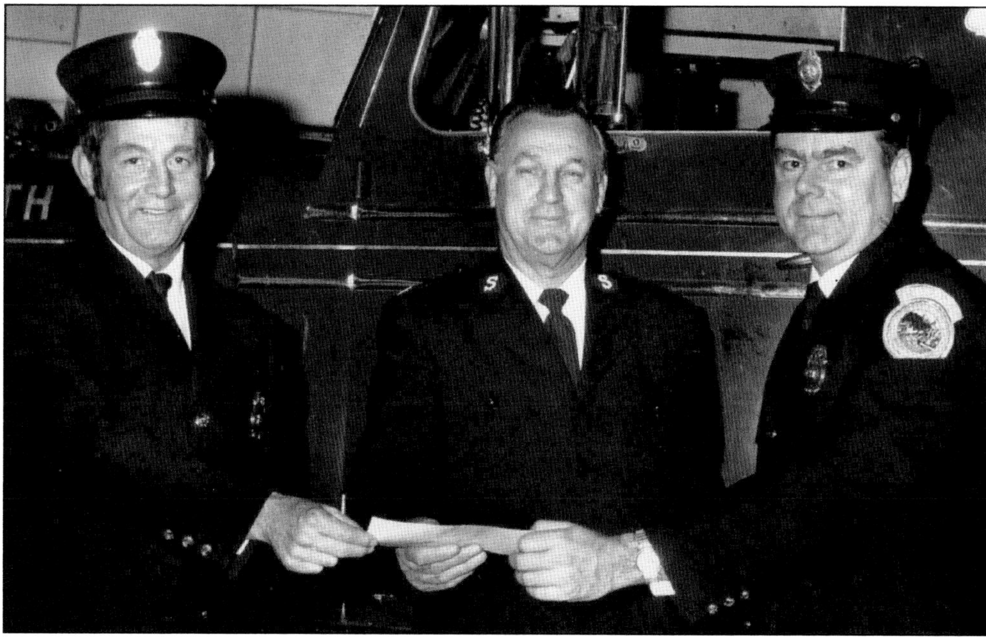

The Salvation Army had a history of providing aid to the firefighters of the department during large fires. In this photograph taken in November 1977, in which members of the Portsmouth Firefighters Relief Association are donating funds to the Salvation Army for a canteen truck, are, from left to right, firefighter Frank Collins, Maj. Jack Ryans of the Salvation Army, and firefighter Bill Horvath. Major Ryans also provided chaplain services to the department.

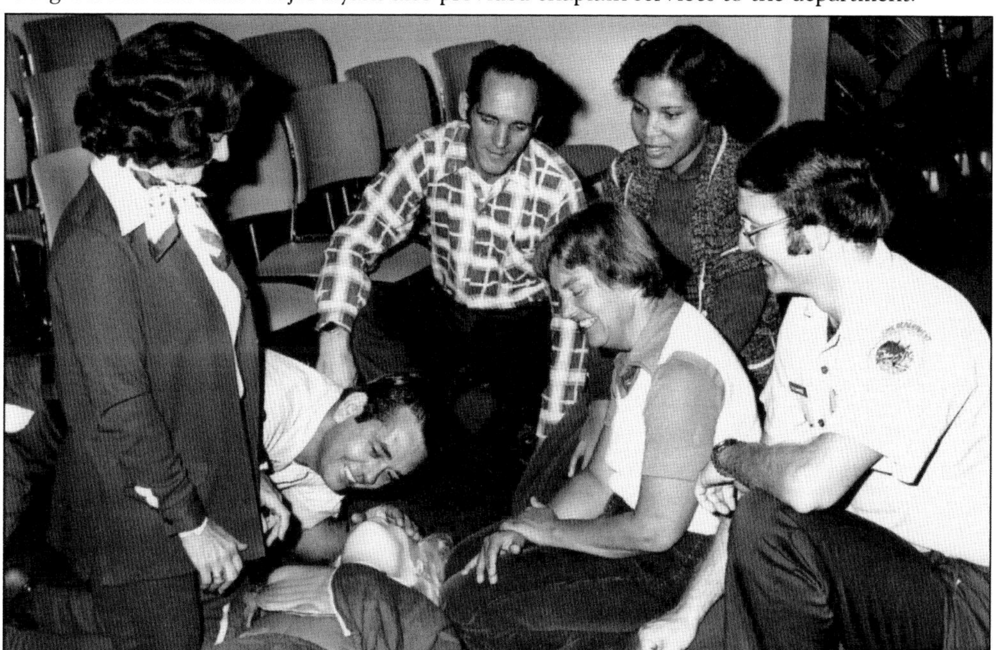

Members of the department provide a CPR demonstration to the staff of the Portsmouth Library in this November 1978 photograph. Firefighter Everett "Al" Weare (left) simulates checking the breathing of a rescue manikin while firefighter Ted Boisvert (right) looks on.

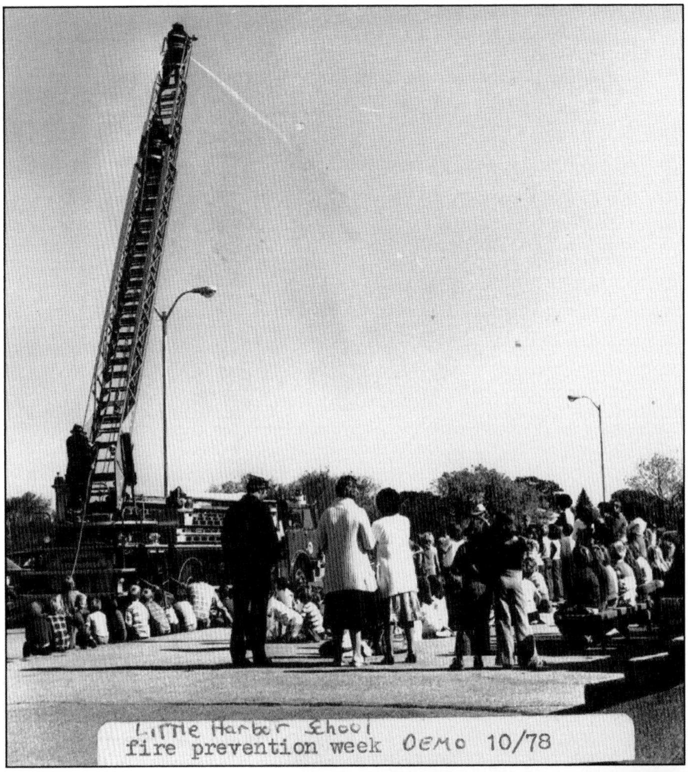

Aerial master stream operations are demonstrated to students of the Little Harbor Elementary School in Portsmouth during Fire Prevention Week in this photograph of ladder No. 1 and its crew taken in October 1978. Fire Prevention Week was established to commemorate the great Chicago fire of 1871. In 1920, Pres. Woodrow Wilson issued the first National Fire Prevention Day proclamation, and since 1922, Fire Prevention Week has been observed on the Sunday through Saturday period in which October 9 falls.

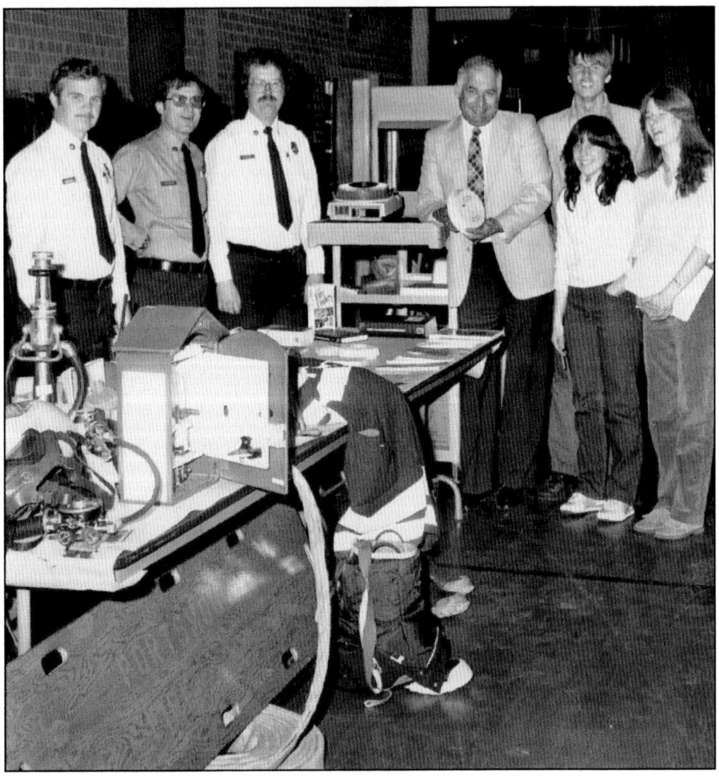

In 1977, Chief Paul Long established the Bureau of Fire Prevention and Control at the Portsmouth Fire Department. George Pierce was hired as the deputy chief of the bureau and served in that role until 2000. In this photograph taken in 1978 are, from left to right, Deputy Chief Leonard Goyette, Capt. Ralph DiBenardo, and Deputy Chief Pierce as they display the "tools of the trade" to principal William Mayo (holding the smoke detector) and three students at Portsmouth High School's 1978 Career Day.

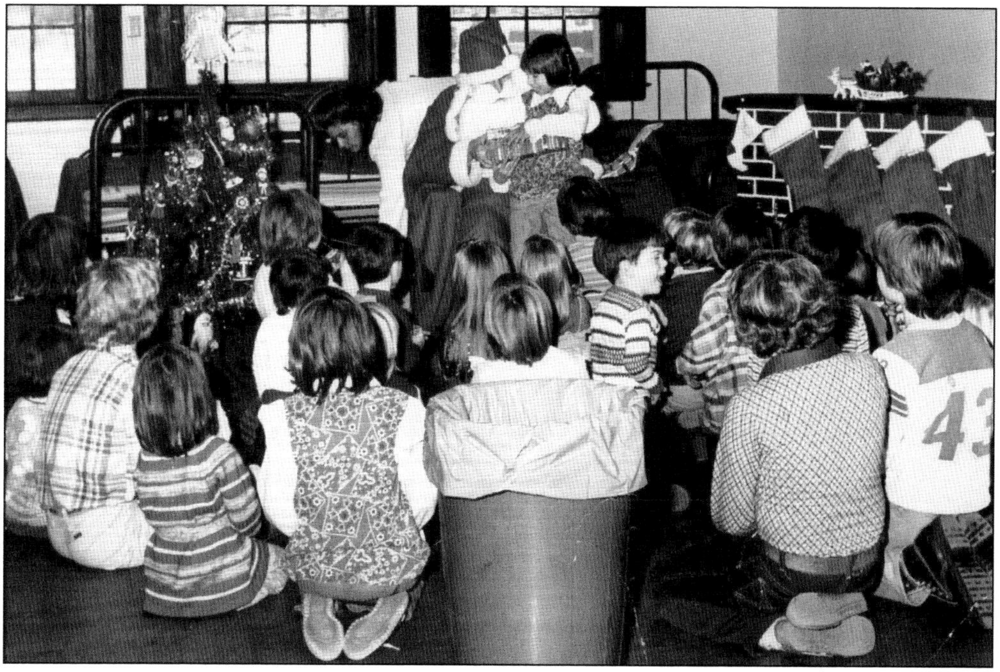

Children and family of Portsmouth firefighters gather around Santa Claus while his helper Marilyn Boone reaches for another gift to be given out in this wonderful image of the department's 1978 Christmas party. Playing the part of Santa Claus is firefighter Herald F. Good Jr., who was a member of the department from 1974 to 1982. The photograph was taken in the great hall at the Court Street firehouse.

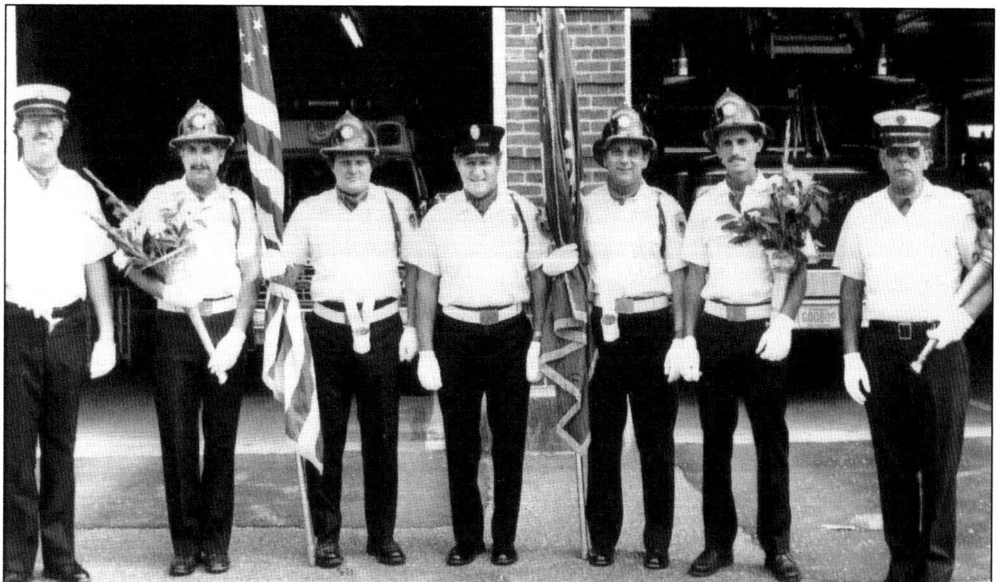

Members of the Portsmouth Honor Guard stand in front of the Central Fire Station in preparation for the Memorial Day parade in this image from the 1980s. Seen here are, from left to right, Deputy George Pierce, Dan Hovey, James Coughenour, Tom McComb, Paul Ward, Charles Cormier, and Lt. Everett "Al" Weare.

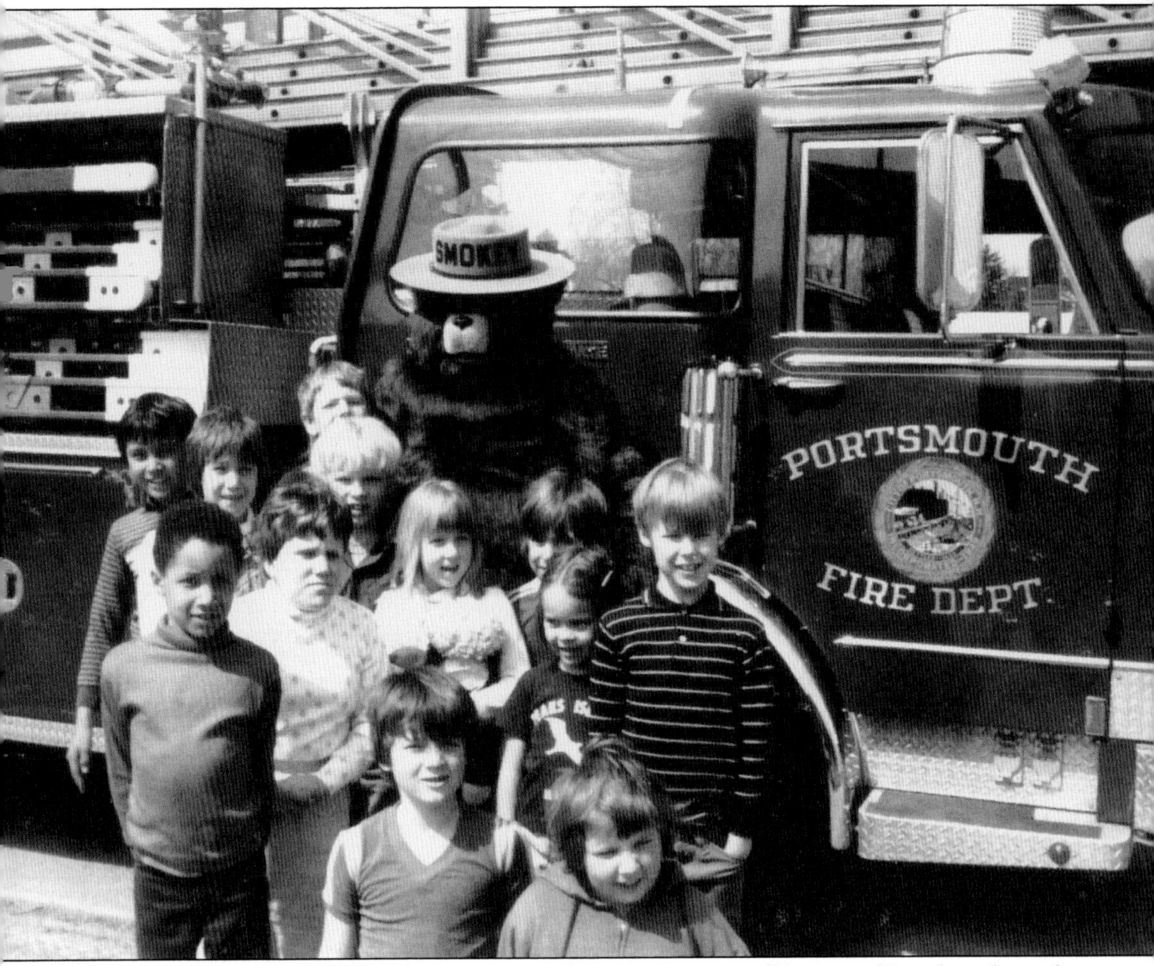

District Chief Al Grimes of the New Hampshire Department of Forest and Lands and members of the Portsmouth Fire Department visited the students of Little Harbor School on April 4, 1984. A film presentation was shown, and Smokey the Bear (Al Grimes) gave a talk on forest fire prevention. Shown here next to Portsmouth Ladder No. 1 are Smokey, James Perril, Allen Fretwell, Susan Jackson, Matt Brewster, Frank Ott, Lisa Scarzik, Amelia Davis, Shane Elliot, Elizabeth Donahue, Geramy Wiason, Seth Gowin, and Tammy Springer. Smokey the Bear and his message, "Only You Can Prevent Forest Fires," are a very familiar and important public service campaign that was introduced on August 9, 1944, to raise awareness to protect American forests.

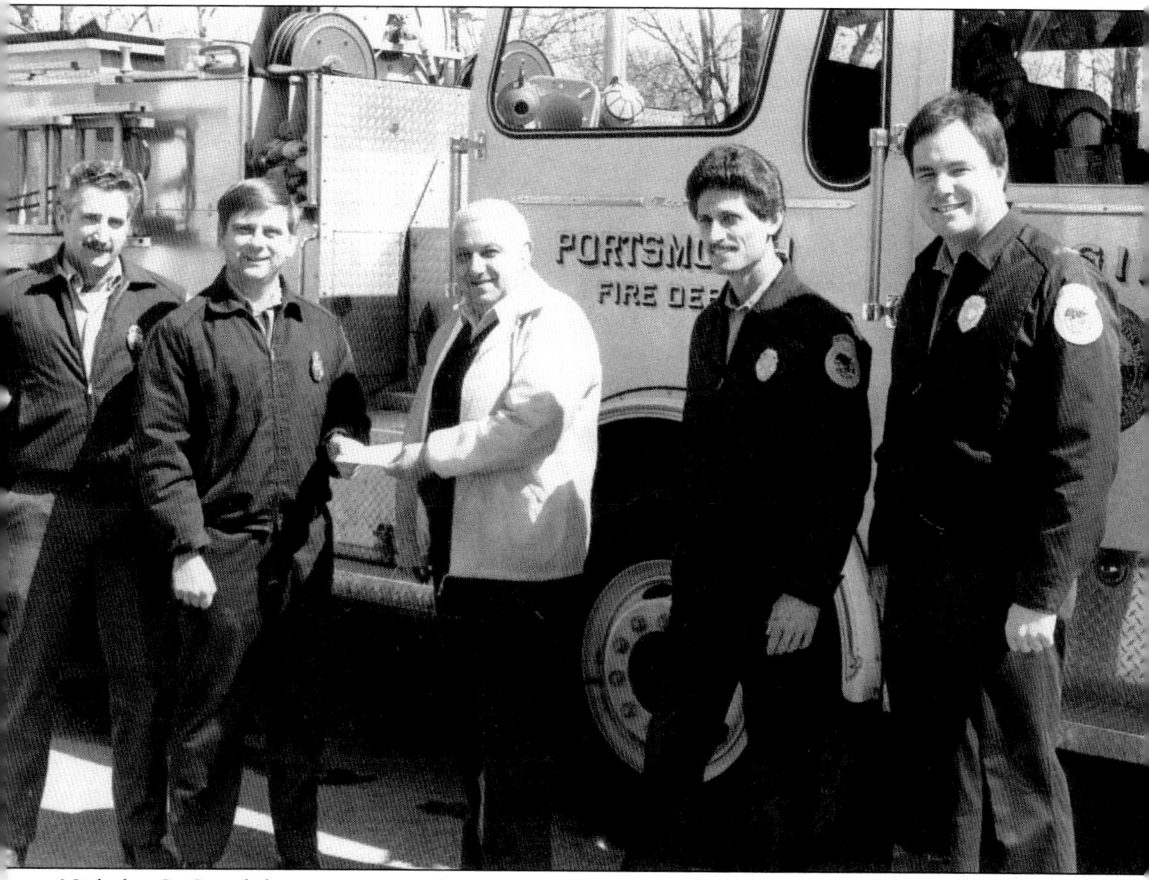

Nickolas C. Grimbilas was appointed to the department as full-time firefighter on July 11, 1959, and on July 1, 1962, he became a permanent firefighter. On April 9, 1983, firefighter Grimbilas was stricken by a heart attack and died while playing basketball at the Connie Bean Center in Portsmouth. The Grimbilas Fund was created in his honor to support youth athletics in the city. The members of the department raised money to donate to the Grimbilas Fund and in April 1985 presented a check to Nick's brother Peter. In this photograph taken during the presentation are, from left to right, firefighter Dan Hovey, firefighter Paul Ward, Peter Grimbilas, firefighter Charles Cormier, and Lt. Steven Griswold.

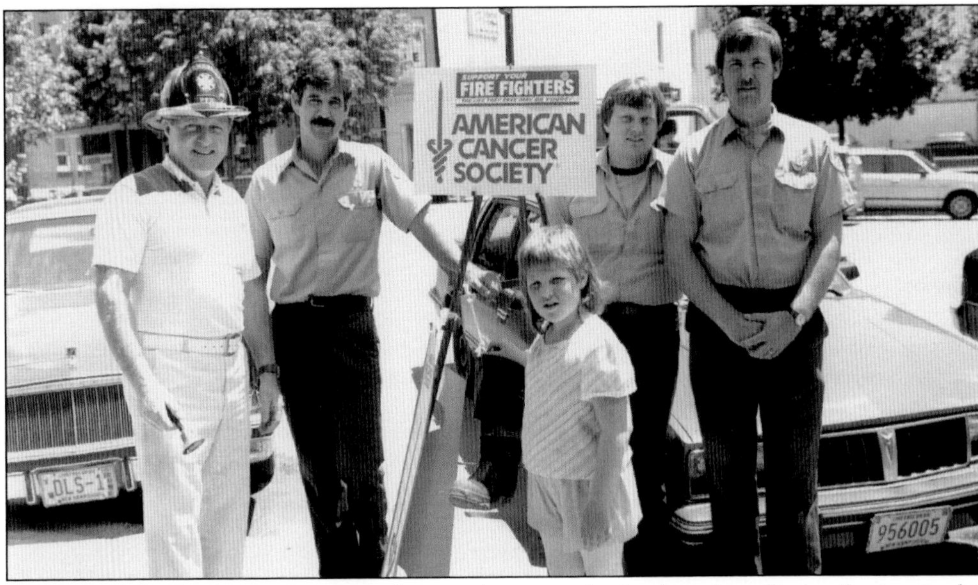

Members of the fire department are collecting donations for the American Cancer Society in this June 7, 1985, photograph taken on Congress Street. Portsmouth firefighters in the picture are, from left to right, Tom McCoomb, Val Pamboukes, James Coughenour, and Charles "Charlie" Jones. The young girl is unidentified.

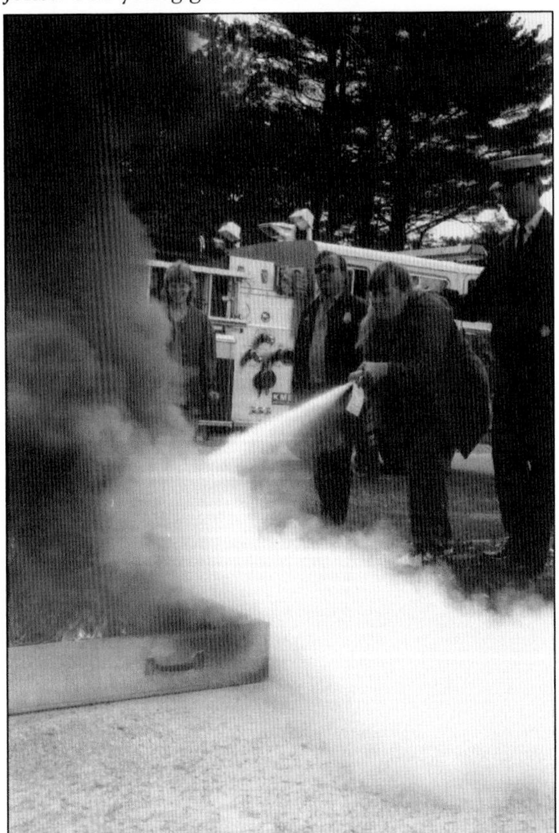

Members of the Portsmouth Fire Department participate in Portsmouth Hospital's 1988 Safety Fair by having hospital staff extinguish small grease and oil fires. In the photograph from June 2, 1988, are, from left to right, unidentified, firefighter Michael Varney, an unidentified Portsmouth Hospital employee operating the extinguisher, and Deputy Chief George Pierce.

BIBLIOGRAPHY

Brighton, Ray. *They Came to Fish*. Portsmouth: Portsmouth 350 Inc., 1973.

Grossman, Nancy W. *The Placenames of Portsmouth*. Portsmouth: Back Channel Press Publisher, 2005.

Ingmire, Bruce E. "Historic 'Kearsarge' fire engine helped save Boston church." *Portsmouth Press*, November 9, 1990.

Murdock, Harold. *Letters Written by a Gentleman in Boston to His Friend in Paris Describing the Great Fire*. Cambridge: The Riverside Press, 1909.

Portsmouth Times, "In the Old Days, How Portsmouth Has Fought Fires," August 15, 1907.

1677-1886 Fires of Portsmouth, Source Material. Works Projects Administration Official Project. Portsmouth Athenaeum Collection.

Sundberg, Joan. *Early Firefighting in Portsmouth, the Development of a Fire Department*. Portsmouth: Interpretative Staff, Strawbery Banke, 1983.

Tufts, Edward R. Letter to Portsmouth Historical Society. Portsmouth Athenaeum Collection.